Nir Arielli is professor of in[ternational history at the University] of Leeds. He is the author o[f ... editor] of *Foreign War Volunteers* and *Fascist Italy and the Middle East*. [He] has also written contemporary political commentary for *Newsweek*, the *Globe Post*, *Haaretz* and *The Conversation*.

Further praise for *The Dead Sea*:

'With its wealth of archaeological insights, historical depth and environmental warnings, this book demands to be read not only as a scholarly text but as a moral imperative.' *Sri Lanka Guardian*

'A lively retelling of the history of the area.' *Life and Work*

'The demise of the Dead Sea is a man-made disaster that requires cooperation to reverse. Arielli's extensive research on the Dead Sea's religious, environmental, economic and conflict dimensions is a must-read for anyone interested in joining the effort to save this natural wonder.' Gidon Bromberg, Israel Director, EcoPeace Middle East

THE DEAD SEA
A 10,000-Year History

NIR ARIELLI

YALE UNIVERSITY PRESS
NEW HAVEN AND LONDON

Copyright © 2024 Nir Arielli

First published in paperback in 2025

All rights reserved. This book may not be reproduced in whole or in part, in any form (beyond that copying permitted by Sections 107 and 108 of the U.S. Copyright Law and except by reviewers for the public press) without written permission from the publishers.

All reasonable efforts have been made to provide accurate sources for all images that appear in this book. Any discrepancies or omissions will be rectified in future editions.

For information about this and other Yale University Press publications, please contact:
U.S. Office: sales.press@yale.edu   yalebooks.com
Europe Office: sales@yaleup.co.uk   yalebooks.co.uk

Set in Adobe Cason Pro by IDSUK (DataConnection) Ltd
Printed in Denmark By Nørhaven A/S, Viborg

Library of Congress Control Number: 2024946019

A catalogue record for this book is available from the British Library.
Authorized Representative in the EU: Easy Access System Europe, Mustamäe tee 50, 10621 Tallinn, Estonia, gpsr.requests@easproject.com

ISBN 978-0-300-28687-8

10 9 8 7 6 5 4 3 2 1

In memory of Ronnie Ellenblum

CONTENTS

List of Illustrations — ix
List of Maps — xi
List of Abbreviations — xii
Note on the Text — xiii

Introduction: A Place of Many Contradictions — 1
1 A Harsh but Welcoming Site — 9
2 A Site of God — 37
3 A Site of Wealth, War and Refuge — 69
4 A Site of Decline, New Beginnings and Myths — 105
5 A Site of Science, Exploration and Diverging Norms — 131
6 A Site of Extraction, Conflict and Water Scarcity — 163
7 A Site of Ruin, Beauty, Health and Hope — 199
Epilogue: A Hitchhiker's Guide to the Beleaguered Dead Sea — 223

Acknowledgements — 231
Timeline — 235
Notes — 241
Select Bibliography — 269
Index — 279

ILLUSTRATIONS

1. The surviving sections of the Madaba map, sixth-century CE.
2. A section of the map *A Pisgah Sight of Palestine*, by Thomas Fuller, 1650. Reproduced with the permission of Special Collections, Leeds University Library.
3. Tell es-Sultan, first decade of the twentieth century. Reproduced with the permission of the Vorderasiatisches Museum in Berlin.
4. Excavations at Tell es-Sultan, early twentieth century. Reproduced with the permission of the Vorderasiatisches Museum in Berlin.
5. Qasr al-Abd. Photo by Vanja Celebicic.
6. Fortress of Machaerus. Photo by Vanja Celebicic.
7. The eastern side of Masada, early twentieth century. Papers of John D. Whiting, Library of Congress.
8. The palm tree 'Methuselah'. Photo by Vanja Celebicic.
9. Monastery of the Temptation. Photo by Vanja Celebicic.
10. The tree of life mosaic, Hisham's Palace. Photo by Vanja Celebicic.
11. *The Scapegoat*, engraved by Charles Henry Mottram, after William Holman Hunt, 1861. The Met Museum (2021.15.19).

ILLUSTRATIONS

12. The north shore of the Dead Sea and Rujum el-Bahr, late nineteenth century. Reproduced with the permission of the Palestine Exploration Fund.
13. Turkish trenches on the shores of the Dead Sea, 1918. Papers of John D. Whiting, Library of Congress.
14. Aerial views of Palestine, at the north end of the Dead Sea, 1930s. G. Eric and Edith Matson Photograph Collection, Library of Congress.
15. The traditional baptism site on the Jordan River. Photo by Vanja Celebicic.
16. Part of the remains of the Byzantine-era complex of churches at Bethany Beyond the Jordan. Photo by Vanja Celebicic.
17. A rock formation on the cliffs east of the Dead Sea. Photo by Vanja Celebicic.
18. Salt accumulation in an evaporation pool near Ein Bokek. Photo by Vanja Celebicic.
19. Brown floodwater spreading across the Dead Sea. Photo by Vanja Celebicic.
20. The terraces left by the receding shoreline on the eastern side of the Dead Sea. Photo by Vanja Celebicic.
21. A westward view from the salt-coated shoreline. Photo by Vanja Celebicic.
22. A sinkhole created by the receding shoreline. Photo by Dr Maor Kohn.
23. The southern basin of the Dead Sea in the late 1960s. Center for Advanced Spatial Technologies, University of Arkansas / US Geological Survey.
24. A contemporary satellite image of the southern basin. Copernicus Sentinel data, 2024.

MAPS

Map 1.	Sites near the Dead Sea from the Neolithic to the Bronze Age.	8
Map 2.	The Dead Sea in antiquity: Settlements, forts and sites of religious significance.	36
Map 3.	Sites near the Dead Sea from the seventh to the nineteenth century CE.	104
Map 4.	The contemporary Dead Sea.	162

ABBREVIATIONS

EBA	Early Bronze Age
ICL	Israel Chemicals Ltd
IDF	Israel Defense Forces
MCM	Million cubic metres
PEF	Palestine Exploration Fund
PLO	Palestine Liberation Organization
PPNA	Pre-Pottery Neolithic A
PPNB	Pre-Pottery Neolithic B
UN	United Nations
UNESCO	United Nations Educational, Scientific and Cultural Organization
UNRWA	United Nations Relief and Works Agency
UNSCOP	United Nations Special Committee on Palestine

NOTE ON THE TEXT

MANY SITES AND some of the groups living around the Dead Sea have had multiple names. Even the same names are often spelled differently in works and studies on the area. The transliteration of Arabic names in particular is notoriously inconsistent. In this book I have opted for the most established and frequently used names. For ancient sites, the most established names are usually those that archaeologists used when surveys and excavations began in the nineteenth or twentieth century. Where appropriate, I provide an alternative spelling when a name is mentioned for the first time.

INTRODUCTION: A PLACE OF MANY CONTRADICTIONS

NEAR ITS EASTERN shore, the bottom of the Dead Sea descends sharply. The sunlight exposes shades of blue and green in the deep water. Guides bring tourists to see a rock formation on the cliff above which, they say, is none other than the pillar of salt that was once the wife of biblical Lot. As the clouds move in from the west, the water and the surrounding hills change colour. Wherever the water touches the shore, the whitened rocks, coated with salt, serve as a reminder: this is no ordinary lake.

The story of the Dead Sea involves salt and sulphur, but also date palms and sugarcane. It can be pieced together thanks to sediments that gathered at the bottom of the lake, scrolls that were hidden in caves, the remains of mosaics and the accounts of travel writers. Its protagonists are Jews and Arabs but also Moabites, Nabataeans, Greeks, Idumaeans, Romans, Crusaders and Mamluks. Over the centuries, the shores of the lake have seen farmers and nomads, biblical figures and pilgrims, conquerors and explorers, great architects and entrepreneurs. Around the Dead Sea one can still find the remains of ancient cities, some of the oldest the world has seen, fortresses perched on hilltops and caves where rebels and refugees sought safety. It is a place that could be many different things – indeed, many opposing things. At different times, the Dead Sea area was seen as both hostile and inviting, sacred and cursed, a seat of war and a resource that could bring prosperity. The lake is famed for having healing properties but, if ingested, its water can kill.

Though it is called a sea in many languages, it is in fact a medium-sized inland body of water, measuring about 76 kilometres from north to south and between 10 and 17 kilometres from east to west. The lake is trapped in a very deep basin created by a geological rift that is still active. The shores around the Dead Sea are the lowest place on land, anywhere on Earth. Water flows into the lake, primarily through the Jordan River, but does not leave it except through evaporation. The salt content of the lake's water is so high that, throughout most of history, it was believed to be completely devoid of any life. As Aristotle observed in the fourth century BCE, 'salt water weighs more than fresh', and it is this salinity that gives the Dead Sea its famous buoyancy.[1]

While working on this book, I asked several friends what is the first thing they think of when the Dead Sea is mentioned. Some said the Dead Sea Scrolls. Others recalled images of tourists floating on their backs while reading a newspaper. Others still thought of the Dead Sea drying up as it runs out of water. It is a peculiar place with a multitude of stories to tell. To weave these together, I took inspiration from a book I thoroughly enjoyed reading many years ago, *The Bridge Over the Drina* by the Yugoslav author Ivo Andrić. The protagonists of the novel are a bridge and the town that grows next to it. While these remain at the centre, people and historical events come in and out of focus. In a nod to Andrić, the protagonists of this book are sites in the area between ancient Jericho in the north and biblical Zoar – modern-day Ghor es-Safi – in the south. Occasionally, we will venture further afield in the southern Levant – the region today comprising Israel, Jordan and the Palestinian Territories. But most of our time will be spent in the immediate vicinity of the lake.

INTRODUCTION

The histories of various bodies of water have sparked a lively debate about the relationship between people and nature and which of the two has been more significant in shaping the course of events. Fernand Braudel's influential *The Mediterranean and the Mediterranean World in the Age of Philip II*, first published in 1949, argues that human history was determined by geography. Geographical features were key to the development of specific economic and state systems, while human endeavours and events are nothing but crests of foam that the tides of history carry on their surface. According to Braudel's approach, people are imprisoned in a destiny in which their agency is limited.[2] In *The Great Sea*, a more recent history of the Mediterranean, David Abulafia seeks to argue with Braudel, advancing the notion that human hands spin the wheel of history. He accepts that geography played a large role in setting the scene, but his focus is on human decisions: where to construct a port, even in places where physical features seemed unfavourable, when to wage war, and with whom to trade.[3] A similar approach is used in other histories of the Atlantic, Indian and Pacific oceans, emphasising how these have been sites of commerce, communication and conflict, and how they enabled the movement of people, goods and ideas.[4]

Of course, the Dead Sea is much smaller. The relationship between people and the environment around it was governed by a particular set of changing conditions. Where settlements were established, to what extent they thrived, why they declined or were abandoned and indeed much of the human history of the Dead Sea was shaped by the interaction between three things: changes in the local environment (both natural and human-made), shifting economic opportunities and geopolitical developments.

Because of its climate and topography, the Dead Sea area has never been densely populated. Nowadays, the area north of the lake receives an average annual rainfall of about 100mm. As one moves south, this amount decreases, falling to approximately

40mm near its southern edge. Surrounded by desert, temperatures can get very high, especially in the summer. In July 2019, a temperature of 49.9°C was measured south of the Dead Sea – the highest since records began in the early 1940s. As a closed – or terminal – lake with no outlet, it has always been highly sensitive to changes in rainfall and river flow. Rainfall in the large catchment area that drains into the Dead Sea changed repeatedly throughout history, meaning that the level of the lake periodically rose and fell. Furthermore, the entire area around it was, and is still, prone to earthquakes. The lake's sedimentary record as well as written historical accounts illustrate that the Dead Sea and Jordan Valley region usually sees one large earthquake or series of earthquakes per century.

Despite these challenges, since the end of the last Ice Age and the beginning of the Holocene, the Dead Sea tells a story of human resilience, resourcefulness and adaptation. Through the ages, human communities in the area found ways to harness local springs, make use of the natural resources provided by the lake and experiment with growing different crops. Indeed, the Dead Sea area offered several economic opportunities. Some of these persisted for centuries while others changed with time because of shifts in demand and the ability to transport and trade goods. Opportunities included growing specific crops that were suited to the hot climate and benefited from the rich soil in nearby oases, and extracting salt, minerals and other substances that could be harvested from the lake itself.

In recent decades, human intervention has increasingly been reshaping the environment of the lake and its surroundings. The building of dams and creation of diversions in the rivers and streams that flow into it have caused the Dead Sea to shrink in size. As is the case with other terminal lakes around the world, the sharp fall in the amount of water flowing in has created an environmental crisis. With the water level of the Dead Sea dropping by more than 40 metres since the mid-1970s, a new landscape has

INTRODUCTION

emerged by its shores. Indeed, the receding lake has created new challenges for communities in the vicinity.

As well as having been shaped by changes in the environment and in economic opportunities, the history of the Dead Sea has also been the product of the shifting tides of geopolitics. Today, the lake is politically divided: an international border passes through it, and Jordanian, Israeli and Palestinian communities live by its shores. This is in keeping with long phases in its history: the area around the Dead Sea has long been home to multi-ethnic and religiously diverse communities. While settlements could have a long lifespan, the political entities that controlled the area changed multiple times through the ages. The land west of the lake has been known as Canaan, Judah, Judea, Palestine and Israel, among other names, while the one to the east was called Moab and Edom, Perea, Oultrejordain, Transjordan and the Kingdom of Jordan. However, the Dead Sea was not always a border zone. For long periods under the Hasmoneans, Romans, Byzantines, Umayyads, Abbasids, Crusaders, Mamluks and Ottomans, the lake was not divided but part of one kingdom or empire.

Fortunately for those who want to study its past, the Dead Sea's unique characteristics mean that it acts like a time capsule. Because it is a terminal lake, the sediments that gather on the lakebed reveal much about the history of the vegetation and, by extension, the climate of its whole catchment area. Furthermore, the exceptionally dry climate around the lake means that organic remains and artefacts that were left behind or hidden in caves in the surrounding hills were preserved for hundreds or even thousands of years.

Alongside these organic and archaeological remains, we also have a large variety of written accounts that shed light on several episodes in the lake's history. The Hebrew Bible mentions the lake several times, most frequently calling it *Yam Ha-Melach* – the Salt Sea. The Dead Sea fascinated Greco-Roman naturalists and geographers and is mentioned in the works of classical historians. During this period, it was often referred to as Lake Asphaltites (the Asphalt Lake) or as *Mare mortuum* (the Dead Sea). In the medieval and early modern periods, the lake was discussed by Islamic, Christian and Jewish travellers, chroniclers and pilgrims. The lake earned yet more names, among them *Bahr Lut*, or Lot's Sea. From the nineteenth century onwards, European and North American explorers, surveyors, geologists, archaeologists and natural scientists set out to study the many mysteries of the lake. The period since the First World War has seen an abundance of source material: press reports, diplomatic documents, development plans, photos, films, satellite footage and much more. All of these sources have informed the writing of this book.

There are different ways of telling the Dead Sea's story. Indeed, this book explores how the same body of water was seen, understood and imagined in very different ways with the passage of time. Because it is a place of contradictions that meant different things to different people, each of the following chapters focuses not only on a specific period but also on diverse facets of the lake and its area. Chapter 1 introduces the geology of the region and considers how the Dead Sea was formed, as well as how archaeologists have reconstructed the story of communities that lived in the lake's vicinity from the Neolithic to the Bronze Age. In Chapter 2 we shift our focus to the religious significance of the Dead Sea area, its role in biblical stories and how, over the centuries, Jews, Christians and Muslims traced these stories in the landscape around the lake. Chapter 3 examines the period between the conquests of Alexander the Great and the heyday of the Byzantine Empire, during which the lake area witnessed unprecedented prosperity but was also the scene of bitter battles. Chapter 4 looks at the history of the Dead Sea during the medieval and early modern periods. Over these centuries, there were notable changes in the economy and demography of the lake area.

INTRODUCTION

In Chapter 5 we retrace attempts in the nineteenth century to study the Dead Sea region, analyse the composition of its water and understand how it was formed. These study expeditions also brought European and North American explorers into contact with local Bedouin tribes who were the primary residents of the lake area. Chapter 6 focuses on the twentieth century, in which the whole region witnessed conflicts over territory as well as disputes over the freshwater resources that sustained the Dead Sea. It is also the period when the exploitation of the lake's minerals reached an unprecedented scale. In Chapter 7 we see how increased human management of water resources in the Dead Sea's catchment area over the last sixty years has had catastrophic implications for the lake. We also examine both the extraction and the health tourism industry that the lake presently supports. The chapter ends with a survey of recent plans that have been put forward to stop the lake from receding any further.

Because of its present environmental crisis, the Dead Sea may benefit from renewed international attention. By highlighting the multitude of historical events that have taken place by its shores, and emphasising the significance this small lake has had in terms of religion, trade, warfare and science, I hope to show readers that the Dead Sea is a unique place which should be saved.

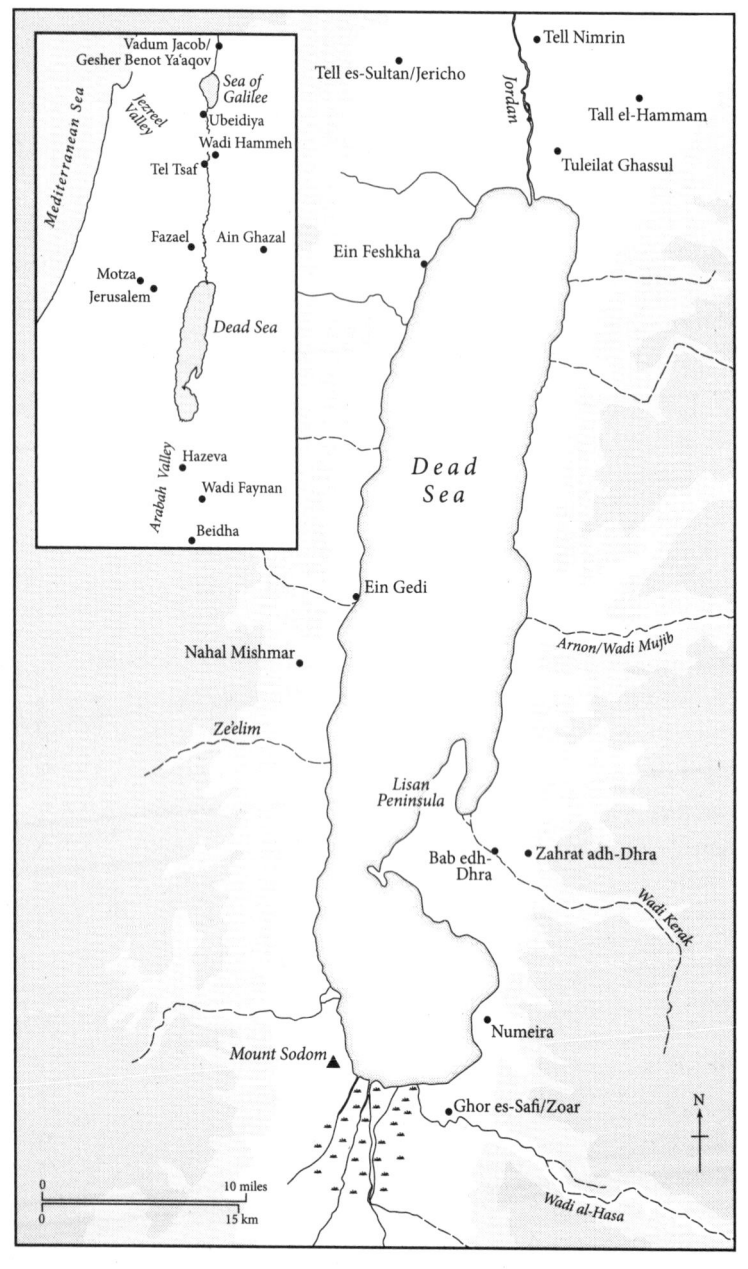

Map 1. Sites near the Dead Sea from the Neolithic to the Bronze Age.

CHAPTER ONE

❧

A HARSH BUT WELCOMING SITE

MANY A VISITOR to the Dead Sea, both past and present, has found the lake and the surrounding barren landscape uninviting. Having travelled extensively through the eastern parts of the Roman Empire, the second-century CE Greek geographer Pausanias noted that the lake has 'properties unlike any other water'. He observed that the lake has no fish and therefore coined the name 'Dead Sea' that lives on in various modern languages. Once they realise the danger they face from entering it, Pausanias explains, the fish escape and 'take refuge in water more congenial to them'.[1] Several centuries later, the Yale geography professor Ellsworth Huntington, who toured the area in 1909, declared that 'The future holds nothing in store for the sea better than the past'. To this he added: 'The hot, un-healthy coasts may in time be visited for their scenery, or for their associations, but the sea is dead, and out of it no life can come.'[2] More recently, the eminent German biblical scholar Wolfgang Zwickel delivered a similarly scathing appraisal: 'The Dead Sea is an unpleasant, waste and extremely hot area, a region where nearly nobody wants to live.'[3] Last, but by no means least, when I asked my five-year-old daughter, Adri, what she thought of the Dead

Sea, she immediately replied, 'I hate it! It's terrible.' Clearly, she had not forgotten the stinging sensation on her skin from her brief dip in the lake nearly two years earlier.

Negative impressions aside, the physical attributes of the region likewise suggest that it would be an unwelcoming site for early human settlement. In the summer, temperatures in the Dead Sea basin regularly rise above 40°C. The annual rainfall of less than 100mm is too low to support rain-fed agriculture. To make matters worse, much of the land surrounding the lake is covered with saline soil which is almost devoid of natural vegetation. Save for some shrubs and other hardy desert plants, the high sodium chloride content of this type of soil impedes the cultivation of crops.[4] Surely the environment of the Dead Sea must have deterred human attempts to settle in the area. Yet archaeologists have shown that, for millennia, it did flourish. How then was such an arid, seemingly inhospitable area one of the earliest sites of permanent human settlement in the world, over 10,000 years ago? To help us answer this question, we must first briefly examine how the Dead Sea was formed.

GEOLOGY, FORMATION AND THE ANCIENT CLIMATE

The history of the Dead Sea, like that of most other places, was heavily influenced – though not altogether determined – by its geology, and the geology of the lake area is unique in many respects. The Dead Sea is a product of plate tectonics. As we shall see later on, the area around the lake gave nineteenth- and early twentieth-century geographers and geologists some hints about the motion of the earth's crust several decades before plate tectonics became a dominant theory in geology.

One useful starting point for the geological story of the origins of the Dead Sea is the gradual disappearance of the Tethys Ocean since the early Miocene between 23 and 16 million years ago. The

African or Nubian Plate and the Arabian Plate have both been pushing northward into the Eurasian Plate for millions of years. Among other things, this tectonic plate movement has produced great mountain ranges such as the Alps, the Taurus and the Zagros. It also severed the long arm that extended from the Tethys to what we now know as the Persian Gulf and Indian Ocean and uplifted the Levant area where our focus lies. What remained of the Tethys – the Paratethys Sea that extended from what is now Romania to Central Asia – was eventually squeezed out of existence around 5.5 million years ago. To the south-west, near the uplifted Levant, there was now an embryonic Mediterranean Sea.[5]

However, parallel to their long, joint movement northward, the Arabian Plate has also been rifting away from the African Plate, a process which opened the Gulf of Aden and the Red Sea. This rifting also created the Sinai Subplate (or Sinai Block) and, in the process, has led to the formation of what we now know as the Jordan Valley. Contrary to popular belief, this tectonic plate movement has not produced one continuous fault line that runs north from the Bay of Aqaba to Lebanon. Instead, it has produced a zone with several faults and basins of varying depth.

Geologists have been debating the tectonic activity underpinning the formation of the valley for several decades.[6] A view that has become dominant since the second half of the twentieth century is that the Arabian Plate, to the east of the valley, has been – and still is – moving northwards faster than the Sinai Subplate to the west, thereby producing seismic activity and rupturing the ground in a way that created the basins of the Jordan River and Dead Sea. Albert Quennell in the 1950s, Raphael Freund in the 1960s and others subsequently demonstrated that geological features on the western side of the valley do not correspond to those directly opposite on the eastern side but have their counterparts far to the north.[7] These observations were later supported by boreholes that were drilled near Masada on the western side of the Dead Sea, revealing a geological stratification the equivalent of which could not be found directly on the opposite side of the lake, but more than 100 kilometres to the north. This suggests that, over millions of years, the eastern Arabian Plate has edged northwards about 105 kilometres compared to its western counterpart.[8] According to this interpretation, the asynchronous movement of the plates has not ceased. My late mentor Ronnie Ellenblum excavated the Crusader castle at Vadum Jacob, north of the Sea of Galilee, in the 1990s and early 2000s. He and his colleagues discovered that the castle's surrounding wall had been ruptured: its eastern part was offset by 2.1 metres compared to the western side of the wall. They ascribed this structural deformation to earthquakes that had struck the region in 1202, 1759 and 1837.[9]

A minority of geologists reject this interpretation, pointing out that, if one plate was sliding past the other, the deformation should be visible continuously, all along the fault line, whereas this is not the case for the central Arabah Valley south of the Dead Sea. Aharon Horowitz, for instance, argues that 'there are at least two sectors of the Jordan Rift where one can cross the valley without stepping on any fault, not even buried'. Instead, he believes that the plates are diverging from one another in a way that is reminiscent of a mid-ocean rift, with a narrow, elongated depression, uplifted belts on each side and pairs of perpendicular faults. According to Horowitz, 'it seems fully justified to regard the Jordan Rift Valley as an embryonic ocean'.[10]

While the exact mechanism of plate movement under the surface in this area remains disputed, there is a broad agreement that, over the course of the last 2 million years, the centre of the valley subsided and the hills to its east and west were uplifted. The two basins which the Dead Sea now covers, the larger and deeper northern basin and the smaller, shallower but older southern basin, are the deepest points in the valley. In fact, they are several hundred metres lower than the present-day level of the Mediterranean Sea.

A HARSH BUT WELCOMING SITE

If plate tectonics account for the formation of the Jordan Valley, the basins under the Dead Sea and the hills to the east and west, how did the lake acquire its hyper salinity? The Dead Sea is the descendant of a series of previous water bodies that differed from each other not only in their age but also in their size and salinity. The late Tethys–early Mediterranean Sea had entered the Levantine Rift Valley during the early stage of its formation. However, the water receded at the very end of the Miocene when the Mediterranean basin to the west largely dried up due to the closure of its connection to the Atlantic Ocean and strong evaporation. Following the Zanclean Flood that refilled the Mediterranean basin with seawater coming from the Atlantic Ocean 5.33 million years ago, the Levantine Rift Valley once again became submerged, with water flowing in from the rising sea through what we now know as the Jezreel and Beisan valleys.[11] The connection between the Mediterranean Sea and Jordan Valley was severed approximately 2 million years ago. This period of connection to the sea, during which the valley was effectively a lagoon, left behind a huge deposit of salt sediments, thousands of metres deep, which accumulated at what was then the lowest area in the valley. Mount Sodom, essentially a huge uplifted structure – or diapir – of rock salt south-west of the Dead Sea, can give contemporary visitors a sense of the massive amount of salt the receding sea left behind it.[12] For comparison, the Mediterranean Evaporite, huge salt deposits that were discovered beneath the sea floor and that formed during the so-called Messinian salinity crisis, a series of desiccation (drying up) events that began some 6 million years ago and ended with the Zanclean Flood, have a thickness of up to 1,600 metres.[13] While the Mediterranean Evaporite covers a much larger area compared with the narrow depression in which the Dead Sea lies, the latter did provide a very efficient lagoon where evaporation accumulated an impressive volume of salts.

After the disconnection of the lagoon from the open sea, the basin was covered by a succession of terminal lakes. Geologists call the middle-to-late-Pleistocene body of water in the area Lake Amora. Its successor, Lake Samra, covered the Dead Sea basin from around 135,000 until 75,000 years ago, leaving behind it deposits about 20 metres thick. The level of this freshwater lake alternated between 380 and 310 metres below the sea level of today, meaning that it was tens of metres higher than the Holocene Dead Sea. These fluctuations seem to correlate with the changes in climate and sea-level rises in the northern hemisphere during the last interglacial period.[14] We know less about Samra than we do about the larger body of water that replaced it.

Between c.70,000 and 15,000 years ago, almost the entire Jordan Valley was covered by the Dead Sea's late Pleistocene precursor, Lake Lisan. Lacustrine sediments, often called Lisan marls, which are visible to this day in various places around the Dead Sea and in parts of the Jordan Valley, give an idea of the extensive area it once covered. The French geologist Louis Lartet, who toured Ottoman Palestine and Transjordan in 1864, was the first to refer to the salt and gypsum-rich sediments he encountered to the south and north of the Dead Sea as the Lisan Formation. He picked the name because he observed that the Lisan Peninsula (Lisan means 'tongue' in Arabic), which in the nineteenth century protruded into the Dead Sea, was for the most part made up of these deposits. Lartet suggested that this tell-tale formation is the remainder of a much larger body of water that preceded the modern Dead Sea.[15]

Lisan was a brackish water lake that supported a variety of marine life. Indeed, Lartet found fossils of *Ostrea* in the cliffs above Ein Gedi on the western side of the Dead Sea and elsewhere around the lake. Lisan benefited from the climatic conditions of the last glacial period. While this period was cold and dry in much of the northern hemisphere, in the southern Levant the climate was cold and wet, owing to the southward migration of storm tracks. Like Lake Samra before it, Lisan's level fluctuated as a result of hydrological changes. In wetter periods such as the one around 25,000 years ago, it reached a maximum height of approximately 170 metres below present-day sea level and stretched for more than 200 kilometres, from Hazeva in the south to the Sea of Galilee in the north.[16]

Then, around 17,000 years ago, an abrupt arid period began. Approximately 14,000 years ago, the lake dropped to around 465 metres below current sea levels, lower even than the Dead Sea's level today. Geologist Mordechai Stein and his colleagues speculated that this arid period coincided with ice and meltwater discharges into the North Atlantic which, among other things, led to a change in the climate of the southern Levant.[17] However, a post-Lisan lake persisted for two further millennia, and its level continued to fluctuate. As archaeologist Eva Kaptijn observed, prehistoric sites of human settlement – to which we will return shortly – from before 13,500 years ago (or 11500 BCE) are not found below -203 metres.[18] Eventually, it was the climate event known as the Younger Dryas that finished off what remained of Lake Lisan. This cool period in the northern hemisphere between approximately 12,800 and 11,700 years ago (or 10800 to 9700 BCE) marked the end of the Pleistocene. In the Levant it created very dry conditions. What was once Lake Lisan shrank dramatically. The level of the water that remained in the basin may have been as low as -700 metres. Modern-day travellers can still find traces of Lake Lisan's retreat west of the Dead Sea, along Route 90. In places where the cliffs are not too steep, there are ancient beach ridges that tell the story of the declining level of the predecessor of the current lake. These are long, terrace-like lines, extending horizontally above the road on the slopes of the Judean Desert.

Ever the contrarian, Horowitz argues that Lake Lisan's retreat was not only the result of changing climatic conditions and evaporation, but also of a relatively abrupt change in the depth of its basin. According to his interpretation, a sharp faulting phase between approximately 18,000 to 14,000 years ago completely changed the landscape of the entire Jordan Rift Valley. He believes that the centre of the northern Dead Sea basin subsided at an incredible rate of 22 metres per 1,000 years. Another basin that subsided around the same time, though not at the same rate and without reaching the same depth, was just north of the Lisan Lake, where the Sea of Galilee is to be found today. Therefore, Lake Lisan did not dry up only due to decreased rainfall and river flows, but also because its waters were drained into a new, deeper basin. As evidence, he points out that, in the central Jordan Valley, the elevation of Lisan Formation sediments, left on the sides of cliffs by the lake's shorelines, can be found at around 180 metres below modern-day sea level. However, these cliffside sediments are found at least 150 metres lower in the wadis (ravines or valleys) that flow to the Dead Sea.[19] Using data derived from drillholes, geologist Yuval Bartov and his colleagues were able to show that the deeper, central part of the Dead Sea basin has been subsiding faster than its margins, though at a much lower rate than Horowitz suggested. At the same time as this subsidence, two massive rock formations, or diapirs, created by ancient salt sediments – the Lisan Peninsula in the east and Mount Sodom in the south-west – have been continuously rising, uplifted by the plate movements beneath them.[20] Tectonic movement beneath the Dead Sea is still ongoing, and, as we shall see, seismic activity has had a profound impact on the lake's human history.

A SITE OF INNOVATION: THE NATUFIAN CULTURE AND NEOLITHIC REVOLUTION

Hominin groups had lived around what is today the Jordan Valley long before the Dead Sea was formed. The site at Ubeidiya, some 100 kilometres north of the Dead Sea, preserved traces of one of the earliest migrations out of Africa of *Homo erectus*, dating back some 1.4 million years. Further north, at Gesher Benot Ya'aqov, excavations revealed a long-standing community that seems to have controlled and exploited fire to cook fish some 780,000 years ago.[21] Both of these sites are found at a higher elevation than the Dead Sea, and their inhabitants almost certainly lived along the shores of the water bodies that preceded it. As noted earlier, before the retreat of Lake Lisan, settling below -203 metres was not an option.

During and shortly after the coldest period of the last Ice Age, between 20,000 and 14,500 years ago, the entire Levant region was cold and underwent an arid phase. Then, as a consequence of slow changes in the Earth's orbit, known as the Milankovitch Cycles,[22] incoming solar radiation in the northern hemisphere increased and ice sheets began to melt. At the end of the Ice Age, temperatures rose and air circulation patterns and storm tracks shifted northwards while precipitation increased. The Sahara region, for instance, saw the beginning of the African Humid Period when it was dotted with lakes. Grassland and trees covered much of North Africa. The Mediterranean Sea level rose, reducing the size of the southern Levant's flat coastal plain. The shifting climate meant that the landscape of this region was not dry, barren and thorny as it appears today. Instead, it gave rise to oak-dominated woodland that offered a higher biomass of food for humans to exploit. These favourable conditions enabled a growth in the Levant's human population and heralded the emergence of a new, distinct group which archaeologists call the Natufian culture.[23]

From around 13,000 years ago (11000 BCE), the growing population forced the Natufians to adapt their way of life to accommodate more people. Whereas previously there had been no sedentary camps in the Levant, carbon-14 dating has shown that from around 12,500 years ago the Natufians began to establish a series of partly sedentary hamlets or 'base camps'. Ofer Bar-Yosef, the doyen of prehistoric archaeology in the Levant, believes the

Natufian culture 'marked a major organizational departure from the old ways of life'.[24] According to another eminent archaeologist, Anna Belfer-Cohen, the Natufians' culture 'constitutes the first deviation from the traditional way of prehistoric living'.[25] Instead of small groups of nomadic hunter-gatherers, the Natufians adopted a more sedentary lifestyle, with larger groups and a more complex level of social organisation that may have included early agriculture. Their subsistence was still based on hunting and gathering, though with a heavy emphasis on plant processing, as evidenced by the ground stone tools and sickle blades they left behind. Other characteristics of the Natufian culture include a strict separation between living quarters and burial grounds, and their semi-circular partially subterranean dwellings of about 3 to 6 meters in diameter. Though perhaps not unique in this respect, the Natufians seem to have dwelled alongside domesticated dogs, as joint dog–human burials suggest. This period also saw the first appearance of another close neighbour of modern humans: the house mouse.[26]

The 'core area' of Natufian sites was initially associated with the oak, almond and pistachio woodland that then dominated the Mediterranean zone and the Galilee in what is today northern Israel. However, findings from the Jordan Valley as well as sites in southern Jordan clearly show that the Natufian culture extended to more arid areas as well.[27] For instance, there was a large Natufian 'base camp' at Wadi Hammeh, east of the Jordan River and west of modern-day Irbid. The inhabitants of this site kept bone pendants similar to the ones discovered at Fazael, more than 50 kilometres away, on the western side of the Jordan Valley, and used rock-hewn bowls similar to the ones found at Wadi Hisban, 70 kilometres away, north-east of the Dead Sea. They also seem to have used basaltic artefacts made from outcrops east of the Dead Sea, some 100 kilometres to the south, even though basalt rocks could be found much more nearby. La Trobe University archae-

A HARSH BUT WELCOMING SITE

ologist Phillip Edwards proposed that the people who lived in the base camp at Wadi Hammeh could have used watercrafts that enabled the long-distance transport of heavy basaltic stone and other ready-made tools from sites across the Jordan Valley. He points to sedimentary formations in the valley that suggest that a smaller post-Lisan lake persisted until after 11000 BCE. The presence of fossilised ostracods embedded in the sandy layers of the sediment reflect the lake's brackish conditions.[28] This was not yet the Dead Sea we know today. Furthermore, springs, some of which have since disappeared, generated freshwater swamps adjacent to the lake, creating a liveable environment for sedentary Natufian communities, and paved the way for early attempts at cultivation.[29]

A small Natufian site at an elevation of between -260 and -250 metres, near to where the Jordan Valley begins to ascend towards the Judean Hills, provides us with an opportunity to introduce one of the protagonists of this book. The spring of Ain es-Sultan is located a few kilometres north-west of the Dead Sea, next to the modern-day town of Jericho in the West Bank. It feeds one of the largest oases in the Levant and has sustained human communities for over 12,000 years, though not without interruption. With the passage of time, consecutive settlements were built at the site, each constructed on top of the collapsed remains of its predecessor, until an artificial mound – *Tell* or *Tall* in Arabic, *Tel* in Hebrew – was formed. Jericho's Tell es-Sultan, which has been at the centre of this long occupation and the scene of so much activity, is perched on a moderately sloping plain, with marls forming its bedrock. These sediments were left behind by the receding lake that regressed from this area around 11000 BCE.[30] The earliest findings uncovered in the layer archaeologists call Sultan 1a date back to about 10500 BCE. These include a scatter of Natufian-style crescent-shaped tools. The residents of the site may have planted cereals, lentils and peas on a small scale.[31]

While Jericho is often described as the oldest city in the world, it was certainly not yet an urban centre at this early stage which was still part of what archaeologists call the Mesolithic. In fact, during the Younger Dryas period, the site was abandoned for nearly a millennium. Between around 10800 and 9700 BCE a massive influx of glacial meltwater into the Atlantic Ocean, caused by the collapse of ice sheets in North America, cooled down the climate of the northern hemisphere. In the Levant, rainfall decreased significantly. Much of the oak-dominated woodland in the area must have died, as the sharp fall in tree pollen grains found in bores suggests. The pressures of the Younger Dryas period forced human groups to abandon many sedentary sites, change their organisational strategies and, in some places, experiment with planting seeds. Wetter conditions returned at the start of the Holocene around 9600 BCE. Annual floods deposited rich fertile soil where wadis opened up into the Jordan Valley. Springs that flowed with new vigour watered these alluvial fans all year round.[32] The subsequent period has been dubbed the 'Neolithic Revolution'.[33] It resulted in the creation of large villages for the first time and heralded the domestication of a number of crops and animals.

It took a while for the extraordinary story of Jericho's Neolithic past to come to light. Because of its biblical association, several scholars sought the remains of the city that Joshua had conquered. Charles Warren of the British Palestine Exploration Fund carried out initial and not very systematic excavations at Tell es-Sultan in 1867 and 1868. The site was also excavated by German archaeologist Carl Watzinger and theologian Ernst Sellin between 1907 and 1909. The Head of the British School of Archaeology in Jerusalem, John Garstang, made some inroads into the Neolithic layers of the Tell in the 1930s. However, the first person to truly uncover the Neolithic village at Jericho was the University of Oxford archaeologist Kathleen Kenyon. Digging for Jericho in the 1950s, she made a number of important discoveries.

A HARSH BUT WELCOMING SITE

The first of these was the birth of architecture at the site. Tenth-millennium-BCE residents of Tell es-Sultan used mudbrick made from readily available material in the area – mud and straw. Early mudbricks were shaped like bread loaves and used to build curvilinear buildings. Homes initially had curved walls, but over time building techniques developed, enabling the construction of rectangular domestic units. In addition to houses, the increasing cultivation of grain required the development of structures for storage. And indeed, agriculture flourished in Jericho's oasis. The Italian-Palestinian archaeological expedition led by Lorenzo Nigro, which has been excavating Tell es-Sultan almost continuously since 1997, has unearthed evidence of the cultivation of barley and emmer wheat, the construction of canals in the oasis and the presence of domesticated sheep, goats and bees.[34] All these innovations were undoubtedly accompanied by new social norms and greater societal cooperation needed for construction projects and larger-scale farming.

While architecture and agriculture changed dramatically, Jericho's archaeological record from the early Neolithic period lacks a quintessential characteristic of later Mediterranean and Western Asian societies: pottery. Already in the early 1950s Kenyon observed that 'the pre-pottery phase at Jericho would seem to be safely earlier than the majority of Neolithic sites'.[35] She later usefully subdivided this period into the Pre-Pottery Neolithic A (PPNA), which lasted from the early tenth to the early ninth millennium BCE, and Pre-Pottery Neolithic B (PPNB) from later in the ninth to the mid-seventh millennium BCE.

For its time, PPNA Jericho was huge. Already in the ninth millennium BCE it hosted a large community of more than 500 people. The settlement increased in size to at least 2,000 individuals by the middle of the eighth millennium. Archaeologists found evidence of a long-distance trade network in which residents of Jericho must have participated. For instance, they uncovered large amounts of obsidian, a very fine, jet-black and shiny volcanic glass originating from a single source in the hills of southern Turkey. The PPNA and PPNB also saw the emergence of new rituals. Kenyon and subsequent archaeologists found several decorated human skulls, with facial features restored in plaster and eyes inset with cockle shells.[36]

But perhaps the most extraordinary feature that Kenyon uncovered in Pre-Pottery Neolithic Jericho was the protective wall and what appeared to be a watchtower that were constructed on the western side of the Tell. Built of stone, the remains of the wall are about 1.8 metres wide and 3.65 metres high, though it may have been as high as 5 metres during the PPNA. The round tower is attached to the wall on the inner side. It had a diameter of 8.5 metres at its base and reached a height of over 8 metres. At the foot of the tower there was a series of silos to store liquid and dry products. Approximately a hundred people working for a hundred days would have been necessary to build the wall and tower. Kenyon assumed these fortifications were built to protect the settlement from its enemies and compared the tower to the ones found in medieval European castles. Because no similar structures had been found from this early period, and such architecture was completely unprecedented in human history, Jericho acquired the reputation of the world's oldest city.[37]

In the 1980s Ofer Bar-Yosef cast doubt on whether Jericho's wall and tower served as fortifications against human attack. 'Who were the enemies of Jericho that justified this communal effort?', he asked. 'Why is there no record of other fortified sites in the Near East either at the time or thereafter up to about 5500 B.C.?'[38] He pointed out that the smaller PPNA settlement that was discovered a few kilometres north of Jericho at Netiv Hagdud did not have similar defences. Bar-Yosef argued that the walls were not built to keep out human enemies, but were constructed in stages as a defence system against floods and mudflows, considering the proximity of two wadis that open up to the plain nearby. He postulated that the tower may have served as a place of communal ritual activities. The idea that the tower served symbolic and ritualistic purposes was later developed by Ran Barkai and Roy Liran. They pointed out that the tower was perfectly aligned with Mount Quruntul (Mount of Temptation, to which we will return later on) in the west, as well as with the midsummer sunset, which could have been a time of celebration. More recently, Nigro has added further suggestions. The tower could have facilitated hunting, allowing game to be observed at the spring of Ain es-Sultan. The wall could have protected the town and its precious livestock not only from enemy attack but also from theft or natural threats such as wild animals as well as floods.[39]

In recent decades Jericho's position as the 'oldest city' in the world has been contested. First, the term 'city' implies that the settlement showed signs of social differentiation, evidence for which was only found in the remains of later periods. Second, exciting finds at Göbekli Tepe in south-eastern Turkey and in other sites in Western Asia have shown that Pre-Pottery Neolithic settlement was much more widespread than previously believed. Archaeologist Steven Mithen, for instance, suggested in the early 2000s that 'it had been at Göbekli and not Jericho that the history of the world had turned.'[40] However, the point worth emphasising here is that Jericho's importance needn't derive from being the very first 'city' the world has ever seen. Instead, it should be understood as an undoubtedly trailblazing site of residence in the broader Jordan Valley and Dead Sea region which was increasingly dotted with settlements of varying size.

We have no idea what the settlements around the Dead Sea between the tenth and third millennia BCE were called. The name *Ruha*, which most likely refers to the Canaanite-period urban centre we know as Jericho, first appears in the Middle Bronze Age around 1800 BCE.[41] Most of the archaeological sites are therefore identified by their Arabic name at the time when surveys and excavations in the area first began in the late nineteenth and throughout the twentieth century CE: Tuleilat Ghassul, Tall el-Hammam and Tell Nimrin north of the lake, Bab edh-Dhra and Zahrat adh-Dhra east of the Dead Sea and Numeira and Ghor es-Safi to the south-east.

The existence of a PPNA site at Zahrat adh-Dhra was only discovered in the 1990s. It is located on a ridge above the Dhra plain, east of the Lisan Peninsula. Effective harnessing of the nearby spring of Ain Waida and the rainfall gathered by Wadi adh-Dhra from the highlands to the east made life in this arid area possible. Archaeologists found evidence for the cultivation of wild barley and possibly legumes at this small and relatively short-lived settlement. Residents also seem to have collected wild figs and pistachios from the highlands. Like Jericho, Zahrat adh-Dhra was connected to the wider region, as evidenced by the exchange of exotic materials such as marine shells, a handful of flaked obsidian artefacts from Turkey and small beads of green copper ore, which probably came from Wadi Faynan, east of the Arabah Valley and some 20 kilometres south from the edge of the Dead Sea.[42]

By the Pre-Pottery Neolithic B, human settlements in the Levant grew in size and sophistication. The settlement at Ain Ghazal near contemporary Amman was three or four times the size of Jericho. The megasite at Motza, west of contemporary Jerusalem, excavated between 2015 and 2019, had homes with plaster floors that were the product of high-quality construction, requiring considerable community effort in their preparation, including the production of lime. Though smaller, Beidha near Petra in southern Jordan featured two-storey buildings.[43] In the Dead Sea area, chickpea seeds – a staple of Middle Eastern cuisine to this day – made their first appearance in Jericho's archaeological record. The chickpea is drought-resistant and thrives best in a warm climate on moist soils. The residents of Jericho experimented with food storage and preservation, as illustrated by fermented liquids that were found in silos and plastered bins. PPNB sites were also identified near the south-eastern edge of the lake, at the mouth of Wadi al-Hasa, where perennial springs carry rich alluvial soil to the valley floor. These settlements were the forebears of Ghor es-Safi, which in its long history of settlement has also been known as Zoar or Zoara, another protagonist of this book.[44]

In 2021 archaeologists who were looking for much more recent remains made a remarkable discovery that sheds light on everyday life near the Dead Sea in the early PPNB. As part of a last-ditch effort to search caves in the cliffs west of the lake for additional fragments of the Dead Sea Scrolls (to which we will return in Chapter 2), they uncovered a woven basket. Some 50cm high and 60cm in diameter, the basket was found by carbon-dating to be around 10,500 years old (approximately 8500 BCE), making it the oldest ever to be found. In a time before pottery was invented, large and densely woven baskets such as this may have been intended to remain stationary and were used for storage, although it is not yet clear what was kept inside.[45]

Around 6200 BCE the climate in the northern hemisphere became unsettled once again, probably because of yet another collapse of an ice sheet in North America. A period of particularly low temperatures and unreliable rainfall followed. For settlements in the Dead Sea area, the second half of the seventh millennium was one of decline. Jericho degenerated into a small village. The flow of water into the Dead Sea fell, and its level dropped significantly just before 6000 BCE. In such arid conditions human communities were vulnerable. Nonetheless, people continued to use the caves west of the lake even during this period. Findings from some of these caves suggest short-duration occupation in the late seventh millennium as well as throughout the

Neolithic, possibly associated with foraging or herding expeditions.[46] The spirit of innovation would return to the Dead Sea area more than a millennium later, in a period archaeologists call the Chalcolithic.

CHALCOLITHIC ARTEFACTS, FRUIT TREES AND POLLEN ARCHIVES

During the fifth millennium BCE settlement patterns in the Dead Sea region shifted. The settlement at Jericho faced a very severe challenge. Towards the end of the millennium, the spring of Ain es-Sultan reduced its flow or even ceased altogether. As a result, Tell es-Sultan was almost deserted, and a smaller community settled just over two kilometres to the north, closer to two other springs.[47]

While Jericho had declined, on the other side of the Jordan River a settlement situated about 5 kilometres north-east of the Dead Sea flourished from 4500 BCE onwards. Tuleilat Ghassul in contemporary Jordan sits in an alluvial fan at the confluence of two lateral wadis, Wadi Gharaba and the smaller Wadi Ghassul. Hardly visible in the landscape today, the site consists of a number of small Tells that originally stood a little above the surrounding terrain which has since been filled up with alluvium. Excavations at Tuleilat Ghassul began in the 1930s, revealing distinct types of artefacts that soon became associated with a 'Ghassulian culture' that could be traced elsewhere in the southern Levant. Findings included new types of pottery, such as decorated and painted goblets and holemouth jars. Houses at Tuleilat Ghassul had a courtyard with two to four rooms around it. Animal remains indicate the presence of sheep and goats mainly, but also cattle and pigs in smaller numbers.[48]

The site at Tuleilat Ghassul is especially famous for its enigmatic wall paintings. Painted in black, yellow, red, brown and white, these multi-layered frescos were renewed up to twenty times. The best-known among the seven most coherent paintings are 'the star' with its eight rays, 'the procession', which appears to include figures wearing masks, and 'the tiger and the bird'. Uncovered within apparently unexceptional dwelling units, each containing hearths and a variety of storage bins, the seven paintings from Ghassul are unique examples of Chalcolithic mural art in the southern Levant.[49] Art historian Bernadette Drabsch described the paintings as 'unique masterpieces' that include 'architectural features drawn from a bird's eye view, landscape designs, depictions of arguably real-life events, highly individualized masked characters, the first employment of a ground-line, and the use of carefully drafted geometric elements, all demonstrating both technical innovation and extraordinary skills'.[50] She believes the central role of the wall paintings could have been as a social guide and medium of instruction.

Alongside its artistic sophistication, Ghassulian culture benefited from the growing availability of nutritious food. Olive groves are one of the symbols of the Mediterranean region. Wild olives grew in hilly parts of the Levant long before the arrival of *Homo sapiens*, as charred wood evidence from Gesher Benot Ya'aqov dating back almost 800,000 years illustrates.[51] However, it wasn't until the Holocene that humans began to domesticate olives, significantly increasing their yield.

There is a debate among scholars about where precisely olives were domesticated: the north-western Mediterranean, the central Mediterranean or the eastern Mediterranean.[52] As a patriot of the Levant area I am compelled to support the latter. Luckily for me, there is not only archaeological evidence but also indications from fossilised pollen to support this hypothesis.

Archaeologists have found what they see as proof of the cultivation of olives in the Jordan Valley by around 5000 BCE. Dafna Langgut and Yosef Garfinkel examined charred wood remains of olive and fig trees that were uncovered at Tel Tsaf in the central Jordan Valley, west of the river. They point out that the site falls outside the natural distribution area of wild olives and that therefore the trees had to be introduced to it. Olives require at least 400mm of annual rainfall to do well, so it is possible that the trees also benefited from irrigation. Unlike seed and fruit remains, which do not necessarily indicate cultivation in the site environs and may derive from short- or long-distance trade, charred wood remains confirm, in Langgut and Garfinkel's opinion, the emergence of an orchard economy in the Jordan Valley area. With limited means of transportation, the community at Tel Tsaf would have used locally available wood as fuel material. Furthermore, there is evidence of early fruit-tree horticulture in other parts of the Jordan Valley during this period. Charred olive wood remains were discovered in other Late Chalcolithic sites on the eastern side of the Jordan River such as Abu Hamid, and large amounts of olive-pressing waste were found in Pella, across the river from Tel Tsaf. The emerging orchard economy is believed to have increased the socioeconomic complexity of settlements because olives and dry figs were highly suitable for long-distance trade and taxation, which likely led to the accumulation of wealth.[53]

Where does the Dead Sea come into the picture, you might be asking? Well, first of all, charred olive wood remains were discovered in Late Chalcolithic (c.4000 BCE) sites nearer to the lake such as Tuleilat Ghassul. In a natural cave in Wadi el-Makkukh, just north of Jericho, Tamar Schick found the grave of a forty- to fifty-year-old 'warrior' who was buried alongside a broken olive-wood bow and several composite olive wood and reed arrows. Dated to the early fourth millennium BCE, the bow, likely used for hunting, considering the overall lack of weapons from Chalcolithic contexts, had an estimated range of approximately 70 metres.[54]

But another way to detect the intensive cultivation of olives in the Levant is to identify traces of landscape transformation through the palynological record: the study of fossilised plant pollen and spores preserved, in this case, among the sediments that accumulate at the bottom of lakes. The Dead Sea basin, as Frank Neumann and Wolfgang Zwickel point out, 'is an archive of Holocene sediments'.[55] Pollen and dust carried by the wind from the Judean Hills in the west or by the Jordan River from further north accumulated at the bottom of the lake along with other sediments. A sediment core drilled at the western shore of the Dead Sea near Ein Gedi showed that there was a remarkable increase in olive pollen at the onset of the Chalcolithic period, which corresponds with archaeobotanical evidence of olive cultivation in the Jordan Valley. The sudden increase in olive pollen, while other Mediterranean broadleaved trees such as oaks and pistachios remained more or less the same, is further proof that growing olives in the Levant had become much more widespread during this period.[56]

In addition to vegetation dynamics, fossilised pollen from different strata in the Dead Sea basin serves as a proxy record for fluctuations in the region's prehistoric climate. Based on a sediment core drilled near Ein Gedi, Thomas Litt and his colleagues argue that the mid-Holocene – the sixth, fifth and the first half of the fourth millennium BCE – was a period characterised by long intervals with precipitation exceeding present-day levels.[57] However, because different proxy records do not always align with each other, fully reconstructing the paleoclimate of the Levant is notoriously difficult and, 'like bathing in the Dead Sea, the results must be accepted with a sizable portion of salt'.[58] For example, the water level of the Dead Sea, which had dropped just before 6000 BCE, appears to have remained low until the mid-fourth millennium BCE.[59]

Nonetheless, there are various indications that during the Late Chalcolithic period (4600–3600 BCE) and, to a lesser extent, also the subsequent Early Bronze Age the Levant region saw slightly more humid conditions than those of today. The area north of the Dead Sea was relatively densely settled, and the discharge pattern of streams seems to have been different. Eva Kaptijn explains that, instead of the 'quite abrupt and intense flood flows that occur today after rain showers, discharge was more even and less intense (probably due to a different rainfall pattern and especially denser vegetation cover)'.[60]

To improve the odds of survival and to allow settlements to flourish, residents of the Dead Sea area in the Chalcolithic period became increasingly adept at water management. Around the settlement of Dhra, east of the Dead Sea, terrace walls were constructed to limit soil erosion and maximise water use for agriculture as early as 6000 BCE. These walls were at least 0.75 metres high, and, in one case, were at least 20 metres long.[61] Residents of Late Chalcolithic–Early Bronze Age Tuleilat Ghassul made further strides towards domesticating water. As well as extracting water from wells, they channelled rainwater into underground cisterns, which were plastered with lime and covered with stone lids. In addition to providing an early example of plumbing, their buildings were supported by more than one line of stones to prevent rainfall seeping into them.[62]

In addition to Ghassulian wall paintings and pottery, and the expanding domestication of fruit, the Chalcolithic in the Levant had one more central feature. Copper smelting was invented during the Chalcolithic, an age that derives its name from the Greek *khalkós* ('copper') and *lithos* ('stone'). The introduction of metals and alloys marked a fundamental change in the material world of ancient humans, soon affecting almost every aspect of society, from agriculture to warfare to religion. The earliest evidence of copper smelting comes from the Balkans, Anatolia and Iran and dates to as early as the sixth millennium BCE. In the southern Levant metallurgy emerged during the second half of the fifth millennium BCE. Some copper objects used by the 'Ghassulian culture' such as axes are considered 'utilitarian'. These were made by simple casting and relatively pure copper that originated from the nearest ore body at Wadi Faynan, although it was most likely smelted and casted elsewhere. Meanwhile, 'prestige' objects of intricate forms were made by copper-based alloys of varying quantities of arsenic, antimony and nickel brought to the region from long distances.[63]

Once again, it was a cave west of the Dead Sea that yielded the most dramatic discovery when it comes to the use of copper in the Levant during the Chalcolithic. In 1961 archaeologist Pesach Bar Adon uncovered a huge hoard of objects inside what became known as the Cave of Treasures at Nahal Mishmar, some 10 kilometres from Ein Gedi. It included no fewer than 417 copper objects – including 240 mace heads, about 20 adzes and chisels, about 100 sceptres or standards and 10 circular articles labelled as crowns – all tucked away in a crevasse. Findings from the cave's floor included hearths, domestic utensils and a number of burials from the Chalcolithic period, as well as some later remains. There is still some debate about the exact use of the copper objects, many of which are on display at the Israel Museum in Jerusalem. Most scholars believe they served some sort of ritualistic purpose, as, with a few exceptions, they do not seem to have been tools used in daily life. In 1962, excavations were carried out at a temple or a sanctuary, high in the hills near the springs of Ein Gedi. David Ussishkin, who led the excavations at the site, described it as 'an isolated public structure, without any signs of contemporary settlement or activity remains in its immediate vicinity'. A fragment of plaster decoration found at the Ein Gedi sanctuary was similar to that on the painted walls at Tuleilat Ghassul. Ussishkin did not find any objects that could prove that cultic ceremonies took place there. He suggested at the time, and held firm to this interpretation fifty years later, that the objects found in the Nahal Mishmar cave ten kilometres to the south were removed from the Ein Gedi temple for safekeeping.[64] Why the objects needed to be hidden in a hard-to-reach cave remains a mystery.

At the end of the Chalcolithic and beginning of the Early Bronze Age, the Dead Sea area went through another interlude, which could indicate a drier climate phase. The flourishing local culture of the Late Chalcolithic came to an end. The artistic know-how that produced the intricate wall paintings at Ghassul vanished, and, possibly because of changing trade routes, its alloyed copper disappears from the archaeological record. The copper items that appeared in the Early Bronze Age were made differently.[65]

BRONZE AGE HIGHS AND LOWS

The Early Bronze Age saw the rise of renowned ancient civilisations such as the first dynasties of Egypt and the great cities of Mesopotamia. Unlike previous periods, it also yielded written records, though these remain largely silent regarding the Dead Sea area. In the Levant, there are slight discrepancies when it comes to the chronology of transitions between the various phases of the Bronze Age, with some inconsistencies between sites.[66] Subdivisions abound. For our purpose, it will suffice to focus on developments in the Early Bronze Age I (3300 to 3000 BCE), Early Bronze Age II (3000 to 2700 BCE) and Early Bronze Age III (2700 to 2350 BCE), as well as the subsequent Middle Bronze Age (2000 to 1600 BCE) and Late Bronze Age (1600 to 1200 BCE). While trade flourished and there were a number of technological innovations, the Dead Sea area remained susceptible to changes in climate and seismic activity.

At the beginning of the Early Bronze Age I, Jericho's Tell es-Sultan saw the arrival of newcomers bearing a distinct material culture, markedly different from that of the Chalcolithic. They benefited from the renewed flow of the spring of Ain es-Sultan, which had shifted or dried up in the previous period. Gradually, trade links running along the Jordan and

Arabah Valleys and through to Egypt emerged. From around 3200 BCE onwards, Egyptian artefacts make an appearance in Jericho's archaeological record. In 2017, for instance, excavations revealed a cache of five mother-of-pearl shells from the Nile in a room that also contained the remains of eye make-up.[67]

What started as a hamlet and grew into a large village turned, for the first time, into an urban settlement during the Early Bronze Age II. Based on some twenty years of excavations, Lorenzo Nigro describes how, at the beginning of the third millennium BCE, Jericho 'underwent a major transformation with the erection of a unique fortification line consisting of a solid city wall made of dune yellowish brick set upon a two-course foundation of limestone blocks encircling the whole tell. The inhabited area was terraced, divided into quarters by a main street crossing it north-south and, on the hill overlooking the spring and the oasis, a temple was built.'[68] At the same time, pottery-making developed with the introduction of the potter's wheel, and the region's material culture transformed through the use of copper tools and weapons.

Jericho was not the only urban centre in the vicinity of the Dead Sea during the Early Bronze Age II. Bab edh-Dhra, east of the Lisan Peninsula, is best known for its Early Bronze tombs, first excavated by Paul Lapp in the 1960s and subsequently explored and studied by Walter Rast and Thomas Schaub in the 1970s and 1980s. The simple shaft tombs of the Early Bronze Age I evolved during EBA II into mudbrick funerary buildings with inward-sloping walls in the shape of a beehive. By EBA III, burials took place in rectangular mudbrick funerary buildings. The tombs contained not only skeletal remains but also grave goods such as ceramic vessels. Typical objects found in these funerary buildings included knives, axe heads, wooden combs, numerous finely woven linens and cloths and a wide range of beads and shells that were used as jewellery. Grape and peach pits as well as almonds were also found inside sealed tombs. Indeed, the tombs provide some evidence for social stratification, with the largest tomb containing several pieces of gold jewellery.[69]

Early Bronze Age fortifications suggest that settlements in this period were competing with each other. Bab edh-Dhra, with an estimated population of between 600 and 1,000, was fortified with a massive 7-metre-wide wall. Numeira, some 14 kilometres to the south, was also surrounded by a wall and had a tower on its eastern edge. North-east of the Dead Sea, across the Jordan River from Jericho, the settlement at Tall el-Hammam had 5- to 6-metre-wide walls.[70]

As it turns out, the most dangerous foe these fortifications faced were earthquakes. At the end of the Early Bronze Age II, around 2700 BCE, a violent earthquake struck Jericho, badly damaging its walls. The same seismic event severely damaged Tall el-Hammam some 25 kilometres to the east. While other events adversely affected the broader Levant region at this time, including drought and the shifting of Egyptian trade from overland to maritime routes, targeting Byblos and the Syrian-Lebanese coast, the urban centres north of the Dead Sea recovered quickly. Jericho was swiftly rebuilt in the Early Bronze Age III. The defences around the city were doubled, with an outer wall added a few metres away from the rebuilt inner wall. The fortifications were built with new-size reddish bricks and supported with wooden beams of tamarisk and poplar. Tall el-Hammam too was rebuilt following the earthquake, including its 2.2-kilometre-long wall, gates and towers.[71]

At the end of the Early Bronze Age III, around 2300 BCE, Jericho was once again destroyed. Traces of fire, charred bricks and collapsed buildings were found across the site. The inner and outer walls were completely destroyed. Nigro suggested that this destruction was the result of an enemy attack, but we do not have written sources to shed more light on this event. The Tell was

A HARSH BUT WELCOMING SITE

subsequently used as a place of shelter for a few decades before being forsaken entirely.[72] On the south-eastern shores of the Dead Sea, the urban centre at Bab edh-Dhra came to an end during the same period and was later replaced by a smaller village. Numeira, further south, was abandoned altogether.

Jericho or *Ruha* was revived in the Middle Bronze Age after 2000 BCE. The hieroglyphic inscription *Adjmer Ruha*, literally 'administrator of Jericho', was found on a scarab that was placed on the breast of a young woman in a princely tomb in the city's palace. The scarab had been engraved in Tell ed-Dab'a in the Nile Delta area. It provides evidence for a close relationship between the Egyptian kings of the Thirteenth Dynasty and Jericho's local rulers. Finds in tombs from this period also include jewellery and weaponry made of bronze, an alloy of copper and tin that gave the age its name.[73]

Tall el-Hammam seems to have flourished during the first centuries of the Middle Bronze Age. However, it came to a very violent end around 1650 BCE in a conflagration that we will explore in Chapter 2. The following Late Bronze Age, between 1600 and 1200 BCE, was a particularly dark period in the history of the Dead Sea. Settlement activity all around the lake decreased dramatically, leaving a gap in the archaeological record.[74] While the first half of the second millennium BCE saw the water level of the Dead Sea rise to around -370 metres, a height never seen again since, much more arid conditions soon set in. Pollen data from the Late Bronze Age recovered from the western shore of the Dead Sea at Ein Feshkha is characterised by extremely low arboreal vegetation percentages. This period also saw a severe drop of about 50 metres in the water level of the lake, as cores drilled in the Ze'elim Plain, Ein Feshkha and Ein Gedi indicate.[75] By the time climatic conditions improved and the lake level recovered in the Iron Age, the Dead Sea began to acquire a religious significance that still lingers to the present day.

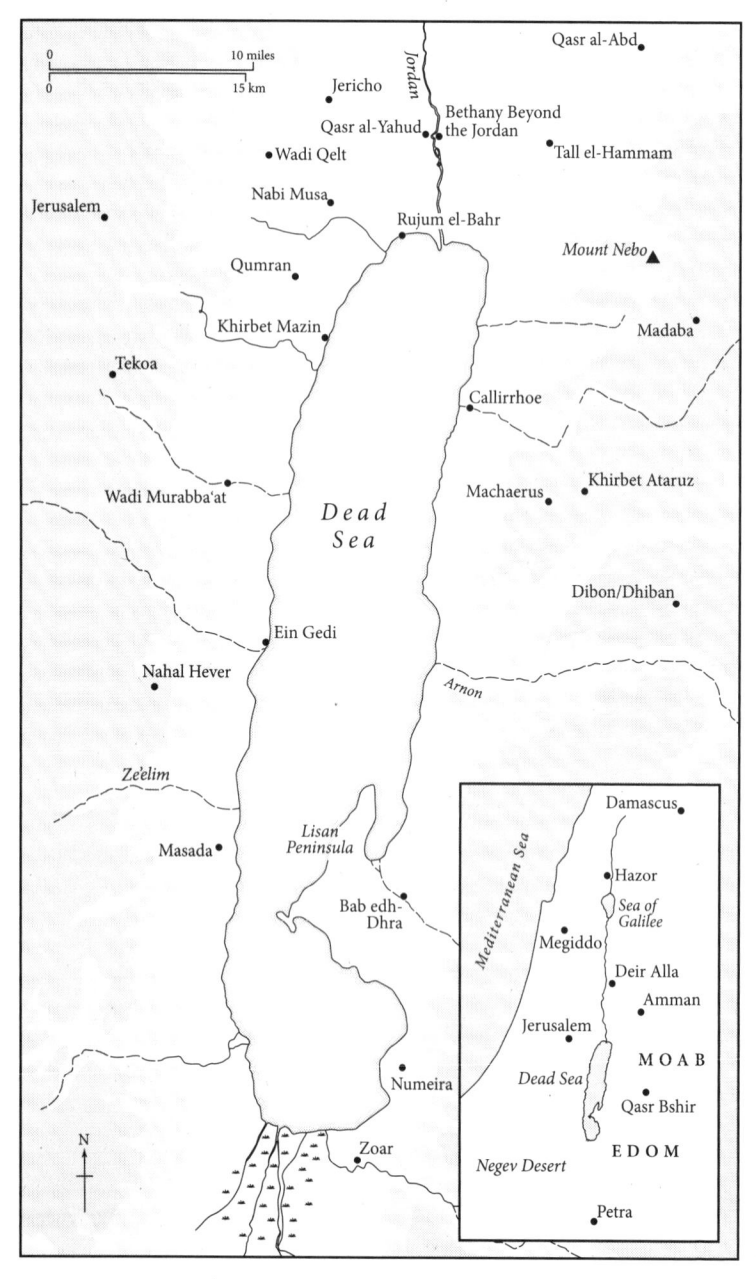

Map 2. The Dead Sea in antiquity: Settlements, forts and sites of religious significance.

CHAPTER TWO

❉

A SITE OF GOD

Qasr al-Yahud is situated on the western banks of the Jordan River, a few kilometres east of Jericho and north of the Dead Sea. For several centuries, the site has been associated with purification. In the Hebrew Bible, the Book of Kings tells the story of Naaman, the commander of the armies of Aram in Syria, who suffered from leprosy. On the suggestion of an enslaved Israelite girl in his household, Naaman came to the prophet Elisha so that he could be cured. Elisha told Naaman to dip seven times in the Jordan River. Though he was reluctant at first, arguing that the kingdom he served in Damascus had better rivers, Naaman eventually did as the prophet said, until 'his flesh was restored and became clean like that of a young boy' (2 Kings 5:14). The text does not say exactly where along the Jordan this baptism – as the first Greek translation of the Bible called it – took place. Nonetheless, Qasr al-Yahud is associated with this story as well as with a later, and probably much better known, purification ceremony: the baptism of Jesus by John the Baptist. According to John 1:28, this latter baptism took place in Bethany Beyond the Jordan, just across the river from Qasr al-Yahud. Excavations in Bethany (also known by its Arabic name, al-Maghtas) began in

1996 and revealed the remains of a complex of fifth- and sixth-century-CE churches, chapels and monasteries. One of these structures included black marble steps leading to a bathing area at what must have been the river level at the time, before the Jordan's course shifted westwards, suggesting the site was already used by pilgrims for baptism in the Byzantine period.

By the turn of the twentieth century, thousands of pilgrims were visiting the banks of the Jordan River every year, especially in January, when Orthodox Christians celebrated Epiphany. In 1906 the American entrepreneur Colonel Clifford E. Nadaud, the President of the International Jordan River Company, received a concession from the Ottoman Sultan to bring water from the traditional baptism site all the way to New York. With the assistance of the local Ottoman authorities, the American Consul in Jerusalem, hoteliers in Jericho and Father Maximos, the Greek Orthodox Superior of the nearby St John the Baptist Monastery, he had fifty-three casks of Jordan water drawn from the river and boiled to disinfect them. The casks were then carried overland through Jerusalem to the Mediterranean port of Jaffa. Back in New York, this precious water was sold in specially designed bottles at $1 each and used, among other things, to baptise babies.[1] As the hundreds of thousands of pilgrims who continue to visit the site every year to this day demonstrate – especially those who dip in the river's far-from-clean waters – Qasr al-Yahud remains a sacred place.

The venerated baptism site is not alone in bearing religious significance. In fact, the area around the Dead Sea features in several biblical stories. Joshua parted the Jordan River east of Jericho to allow the Children of Israel to cross into the promised land at the end of their forty-year journey in the desert. While the Israelites were camped near Jericho at a place called Gilgal, Joshua came back down to the Jordan River where he met the commander of the army of the Lord, who told him: 'Remove the sandals from your feet, for the place where you stand is holy' (Joshua 5:15).

A SITE OF GOD

Joshua was not the last biblical figure to part the Jordan in this area. Several years before he told Naaman how to cure his leprosy, Elisha was but a disciple of the prophet Elijah. In what was to be their last journey together, the two men were travelling eastwards from Jericho. Soon after Elijah parted the Jordan River to ease their crossing, a chariot of fire appeared all of a sudden and the elderly prophet was taken up to heaven by a whirlwind. The area further south, along the Dead Sea coast, also saw a fair share of biblical action. For instance, the young David, before he became King of Israel, found refuge in a cave near Ein Gedi.

For centuries, pilgrims, scholars and travellers who visited the area sought to find the sites where these biblical stories took place. While many associated the Jordan River with purification, the barren and lifeless Dead Sea itself was often described as a place cursed by God. At times, those visiting the lake needed to reconcile the differences between what they saw and the description in the scriptures. As we will see, from the late nineteenth century onwards, archaeology was mobilised in discussions on the veracity of biblical descriptions of the Dead Sea. First of all, let us try to understand what was happening around the Dead Sea where the previous chapter ended, at the end of the Bronze Age.

GAPS IN KNOWLEDGE

The religious significance of the Dead Sea area in Judaism, Christianity and Islam – whether as a cursed site or as a sacred one – stems from events that are believed to have taken place between the Late Bronze Age (1600 to 1200 BCE) and the first century CE. Ironically, the Dead Sea area has revealed little about its own history during the first few centuries of this period. Archaeologists study the remains of material culture. Alas, compared to the preceding Early Bronze Age and the later Hellenistic and Roman periods, archaeological evidence around the Dead Sea from the

Early Iron Age is relatively poor. Before the seventh century BCE, the area was sparsely populated and there seem to have been very few settlements around the lake. Therefore, to understand the social, political and environmental transformations that underpin the religious stories associated with the lake and its environs, we are obliged to rely on evidence from elsewhere in the ancient Near East. Among other sources, we must turn to written texts from Pharaonic Egypt, such as the clay tablet archive known as the Amarna Letters and, of course, the Hebrew Bible, even though parts of the latter may have been composed long after the events it describes.

Textual sources may shed some light on the period's nomadic groups that, because of their lifestyle, are less archaeologically visible. One such group that is believed to have inhabited the area near the Dead Sea during the Late Bronze Age is the 'Shasu'. All we know about the Shasu comes from Egyptian texts and inscriptions from between the fifteenth and twelfth century BCE. From these we learn that the Shasu were nomad pastoralists who were organised in tribal societies and lived at the periphery of Egyptian rule. Though the Dead Sea itself is not mentioned in Egyptian sources, some texts do refer to Shasu presence in southern Transjordan – an area that would later be known as the land of Edom – not far from the lake. The Semitic linguist and historical geographer Anson F. Rainey believed that the ancient Israelites and their eastern neighbours, the Midianites, Moabites and Edomites, may have originated from the Shasu tribes.[2] Over time, these nomad pastoralists adopted a more sedentary way of life and established their own kingdoms.

Although their origin and identity remain under debate, there is little doubt that, after 1200 BCE, new settlers moved into the central hill country west and north-west of the Dead Sea. Beforehand, while cities along the Mediterranean coast of the Levant had flourished during the Late Bronze Age, this hill country remained sparsely settled. Then a huge transformation took place. Around 1200 BCE several kingdoms in the eastern Mediterranean suffered a series of crises. The jury is still out on the precise cause – famine, disease, war, a change in climate that brought about a severe and prolonged drought, mass displacement or a combination of some or all of the above – but the end result was that the prosperous Late Bronze kingdoms of the region collapsed one after the other. The fall of Canaanite city-states such as Hazor, Megiddo and Lachish in the late thirteenth or early twelfth century BCE coincided with this broader period of instability.[3]

The following Early Iron Age saw the appearance of hundreds of new, small settlements at higher altitudes. Irina Russell and Magdel Le Roux suggested that this migration movement coincided with a period of warm and dry climate, since it was easier for people to live in hill country villages than in lowland cities when the weather became drier.[4] By the tenth century BCE these settlements were sufficiently important, or troublesome, to warrant a raid by the Egyptian Pharaoh Sheshonq I – almost certainly the biblical Shishak – who later listed some of them at the Temple of Amun in Karnak. While these hill country settlers gave rise to the kingdoms of Israel and Judah, settlement around the Dead Sea had not yet recovered since the area was largely abandoned at some point in the Middle Bronze Age. As we shall see, some associate this gap in the archaeological record with the biblical story of the destruction of the cities of the plain – a story that inspired a centuries-long search for them that continues to the present day.

FINDING SODOM

If there is one biblical story that the Dead Sea area is associated with most of all it is not one that relates to purity, sacredness or miracles; it is a story pertaining to evil cities of debauchery that were cursed and destroyed. The story of Lot is well known. The nephew of Abraham, Lot had parted ways with the biblical patriarch, setting up his tents and making his home near the city of Sodom. The people of Sodom and neighbouring Gomorrah had acquired a reputation for being evil, and God wanted to destroy them. When two angels appeared at the gates of Sodom, Lot did his best to welcome them, leading them away from the city and into his home. However, all the men of Sodom, young and old, showed up at Lot's door demanding that the two visitors be brought out so they could rape them. Lot refused, offering to bring out his own daughters instead, but the men of Sodom persisted, threatening Lot as well, and moved forward to break down the door. Then, the two angels blinded all the attacking men and told Lot to take his entire household and flee because God would soon destroy the city and all the valley. As Lot, his wife and two daughters were preparing to leave, the angels told him to keep moving and not look back. Soon, the destruction began. Lot's wife, who looked back despite the angels' warning, turned into a pillar of salt. The next day, the author of Genesis adds, the smoke that rose from the valley, as if from a furnace, could be seen as far as the hill country to the west where Abraham lived.

The exact location of Sodom and Gomorrah remained a mystery that intrigued scholars and explorers for several centuries. When describing the geographical location of the cities of the plain, the biblical author used various place names, inconsistently. Some of these place names were not easily identifiable for readers in subsequent periods. Genesis 13:10 says Lot decided to settle in the area because, before the destruction, 'the plain of the Jordan toward Zoar was well watered, like the garden of the Lord'. Later on, however, the ill-fated cities are said to have been located in the Valley of Siddim, which, Genesis clarifies, is the Salt Sea. Subsequent interpretations of the biblical story, such as the one offered by the Roman-era Jewish historian Josephus, explained that the valley became a lake following the destruction of the city of

Sodom.[5] Interestingly, in Genesis, the Valley of Siddim is said to have been full of tar pits, which could become death traps, long before the Lord rained down sulphur and fire upon it. Near Sodom and Gomorrah were three other settlements: Admah, Zeboyim and Bela. Bela, the biblical author explains, was Zoar, presumably a place name contemporaries would have been familiar with. While Admah and Zeboyim were destroyed along with Sodom and Gomorrah, Lot managed to convince the angels to spare Zoar, which was very small, so that he and his family could find refuge there. In the first century CE, Josephus claimed that the shadows of the cities that had been burnt by lightning could still be seen in the area near the Dead Sea.[6]

In the New Testament, Peter mentions that Sodom and Gomorrah had been reduced to ashes as an example of what would happen to the ungodly but does not shed light on where exactly they once stood. The remains of a fifth- and sixth-century-CE monastery, built over a grotto at Ayn Abata on a hillside above contemporary Ghor es-Safi in Jordan, were uncovered in the 1990s. Greek inscriptions found at the monastery identify it as 'Saint Lot's Sanctuary', showing that the site south-east of the Dead Sea was firmly associated with the biblical story in the Byzantine period. From accounts written by pilgrims, it seems there were two rival locations for the pillar of salt that was once Lot's wife: one north and one south of the lake. The late-fourth-century nun Egeria, for instance, was shown by a local bishop where the pillar once stood at the northern edge of the Dead Sea. However, she was told that the pillar had been submerged by the lake's waters.[7]

The Quran gives more prominence to Lot than the Hebrew Bible does. It tells of the destruction of the 'subverted cities' by a rain of stones as a punishment for the wickedness of their residents and even adds that their ruins 'still lie along a known route' (Surah al-Hajir 15:76). However, the Quran does not mention the cities by name nor say where their ruins can be found. Over time, Islamic tradition, much like Judaism and Christianity, began to associate the story of Lot with the Dead Sea. Medieval Islamic authors called the area near the lake 'the Land of Sodom', 'the country of Lot's people' or the 'overturned' land where nothing grows. The medieval settlement near ancient Zoar – Zughar – was said to be named after one of Lot's daughters. *Bahr Lut* – Lot's Sea – became one of the Arabic names of the lake.[8] The mountain of rock slat south-west of the Dead Sea eventually became known as Jebel Usdum – Mount Sodom.

From the Crusader period onwards, some pilgrims and travellers who sought the remains of Sodom and Gomorrah but could not find them suggested that the cities of the plain had sunk in the lake.[9] The maps of Palestine produced by the English preacher, author and historian Thomas Fuller in 1650 show the four destroyed cities, with flames still rising from them, in the middle of the Dead Sea. Nonetheless, the search for Sodom and Gomorrah continued. Lieutenant Claude Conder, who led the Survey of Western Palestine for the British Palestine Exploration Fund (PEF), noted in 1874 that 'having carefully examined in person the whole tract from Jordan mouth to the Ras Feshkah I do not hesitate to say that, if the cities of the plain were within this area, all trace of them has utterly disappeared'.[10] In the 1930s, the American geologist Frederick Gardner Clapp drew on archaeological evidence and his interpretation of the area's geology to argue that 'Most probably the cities of the Plain are buried beneath the waters of the shallow embayment' in the southern basin of the lake.[11]

Archaeologists excavating around the Dead Sea in recent decades have also weighed into the debate about the location of the cities of the plain. Konstantinos D. Politis, who spent several years surveying and excavating in the vicinity of Ghor es-Safi, suggests that Tuleilat Qasr Mousa Hamid could be biblical Zoar, the only one of the cities of the plain that was spared. The site boasts Iron Age II pottery sherds that date to around 900–600

A SITE OF GOD

BCE. Politis also found evidence for substantial farming activity, possibly based on wheat and barley production large enough to sustain a sizeable population.[12]

Walter Rast and Thomas Schaub, who excavated Bab edh-Dhra from the 1970s onwards, found a system of Early Bronze Age towns and bastions which, they argued, 'may bear on the biblical tradition of the "cities of the plain"'.[13] Rast later took the claim even further, arguing that the neighbouring Early Bronze towns of Bab edh-Dhra and Numeira, both bearing signs of destruction and abandonment, may have generated the popular biblical tradition relating to Sodom and Gomorrah. Excavations at Bab edh-Dhra, for instance, revealed four funerary buildings that had extensive burn areas with remnants of burnt roof beams. These structures were damaged at about the same time as the Early Bronze Age III walled settlement at the site came to an end.[14] Alas, the international attention that the discoveries of Rast, Schaub and others elicited also had a negative impact on the study and preservation of ancient Bab edh-Dhra: soon, archaeologists noticed a growth of clandestine activities by local tomb-robbers. Politis suspected that these were 'organised, funded and sometimes armed by regional antiquities dealers closely connected to the international antiquities market'.[15] One wonders how much an association with the notorious city of Sodom adds to the market value of an ancient artefact.

Rast's assertion regarding Bab edh-Dhra and Numeira did not settle the debate regarding the location of Sodom and Gomorrah. Those believing the cities of the plain must have stood north of the Dead Sea remained unconvinced. For instance, Phillip Silvia, an engineer-turned-archaeologist from Trinity Southwest University, insists that Bab edh-Dhra is 'at the wrong place and at the wrong time' to be biblical Sodom.[16] Silvia is part of a large group that has been excavating Tall el-Hammam, a few kilometres north-east of the Dead Sea. In 2021, the Tall el-Hammam Excavation Project made a sensational announcement.

Excavations at the site began in 2006 and have been continuing annually since. The archaeologists found the remains of a Middle Bronze Age city, exceptionally large for its time, which bore remarkable signs of destruction, including 'melted mudbrick fragments, melted roofing clay, melted pottery, ash, charcoal, charred seeds, and burned textiles, all intermixed with pulverized mudbrick'.[17] They also found potsherds that appeared to have melted at high temperatures, but with no clear visible cause. They noted that structures in the southern, south-western and western sectors of the lower city seem to have suffered greater damage than its upper north-east side. Radiocarbon dating suggested the destruction took place around 1650 BCE, give or take fifty years.

What could have caused such wide-ranging damage? A multidisciplinary team of scientists set out to test different hypotheses. They concluded that the destruction found at Tall el-Hammam could not have been caused by an earthquake or warfare, considering the technology available at the time. They also discounted wildfires, as these could not have reached the temperatures needed to melt zircon or shock metamorphism in quartz. Instead, they argued, the unusually high temperature, the scale of damage and its directionality point to a cosmic source: an impact by a meteorite or the nearby explosion in the air of a comet or meteorite, also known as an airburst. As no meteorite crater has been found in the area (though one may be hidden under the surface of the Dead Sea), the team is leaning towards the airburst hypothesis, giving the explosion over Tunguska in Siberia in 1908 as a comparable example. This latter event saw an asteroid explode some 5 to 10 kilometres above the ground, causing damage to an area of approximately 2,000 square kilometres and knocking down an estimated 80 million trees.

The team further speculated that the airburst may have occurred over the Dead Sea, distributing hypersaline water across the lower Jordan Valley. Because they observed a heavy concentration of salt-laden ash in the shallow soil that covers the lower part of the site, they believe Dead Sea water dispersed by the explosion substantially increased the salinity of surface sediments within the city and in the surrounding fields. The destruction wrought by the cosmic airburst, they argue, rendered the area useless for agriculture, therefore accounting for the long gap of between 300 and 500 years in the archaeological record of several sites north of the Dead Sea. This abandonment continued for the entire Late Bronze Age. Importantly for our purpose, the team suggests a possible link to the biblical story of the cities of the plain: 'It is worth speculating that a remarkable catastrophe, such as the destruction of Tall el-Hammam by a cosmic object, may have generated an oral tradition that, after being passed down through many generations, became the source of the written story of biblical Sodom in Genesis.'[18]

This startling hypothesis has yet to win over some of the experts in the field. Lorenzo Nigro, for instance, who has excavated Tell es-Sultan for many years, remains unconvinced.[19] Jericho is some 22 kilometres west of Tall el-Hammam. Both sites are about the same distance from the Dead Sea. Despite the similar proximity and even though Tell es-Sultan is at a lower elevation, archaeologists in Jericho have not encountered the same thick layer of salt-laden ash found at Tall el-Hammam. The debate about the origin of the story of Sodom is likely to continue. Let us now turn to the contribution of the Dead Sea to the Bible beyond the Book of Genesis.

A DRAMATIC SETTING

The Dead Sea area provides the setting for several, often dramatic, stories in subsequent parts of the Hebrew Bible. Here are some of the most prominent. Alongside the stories themselves, our focus will be on attempts in later centuries to locate these places in the landscape surrounding the lake. As a result of traditions that have developed over time, a number of biblical stories are still associated with landmarks in the vicinity of the Dead Sea today.

One key figure who, according to the biblical author, saw the lake but never got to bathe in its waters, was Moses. As the Book of Deuteronomy draws to an end, the ageing prophet who had led the Children of Israel out of Egypt and through the desert on their forty-year-long journey climbs from the plain of Moab to Mount Nebo. From the *Pisgah*, the summit, the Lord showed Moses the whole land across the Jordan as far west as the Mediterranean. He was also shown the 'the valley of Jericho, the city of palm trees, as far as Zoar' (Deuteronomy 34:3), presumably meaning from the northern to the southern edge of the Dead Sea basin. Tragically for Moses, he got to see the land but never crossed over into it, for he died on Mount Nebo. He was buried in Moab 'in the valley opposite Beth Peor, but to this day no one knows where his grave is' (Deuteronomy 34:6).

The site known as Mount Nebo today boasts the remains of a monastic complex that dates back to the Byzantine period. It includes a large monastery that was first established in the fifth century CE, though the oldest pottery sherds found at the site date back to the fourth century. The complex was reconstructed and enlarged in the sixth century, which is when its exquisite mosaic was built. It was still in use in the Umayyad period (661–750 CE) but was abandoned at some point soon thereafter.[20] The site fits the description in Deuteronomy insofar as it sits on a high summit east of the Jordan River 'across from Jericho'. However, while it offers an excellent view of the northern part of the Dead Sea, the Jordan Valley and the hills of the West Bank, one would not be able to see as far north, west or south as Moses did, even on an exceptionally clear day.

Islamic tradition places the tomb of Moses west of the Jordan River, a few kilometres south of Jericho at a site known as Nabi Musa. It is not clear when precisely this tradition began. A hadith

A SITE OF GOD

narrated by Abu Huraira (or Abu Hurayra) says that Moses asked the Lord to let him die at a distance of a stone's throw from the sacred land and that he was buried by the way near an unnamed red hill (Bukhari, vol. 2, book 23, no. 423). Texts from the first few centuries of Islam debated whether the tomb of Moses was located near Jericho or Damascus. By the thirteenth century, there was clearly a tradition that placed the tomb near the road between Jericho and Jerusalem. The Mamluk Sultan Baybars, to whom we will return later on, stopped there on his way back from the Hajj in Mecca in 1269. Baybars ordered the construction of a domed shrine (*maqam*) and, possibly, a mosque at Nabi Musa. The complex was further enlarged to provide facilities for pilgrims in the late fifteenth century, with subsequent repairs and extensions carried out by Ottoman officials in the eighteenth and nineteenth centuries.[21]

After the death of Moses, Joshua led the Israelites across the Jordan near Jericho, leaving the tribes of Gad and Reuben to settle on the river's eastern bank. Following a sojourn at Gilgal, Joshua and his army marched around the walls of Jericho for seven days, eventually making them tumble down by only blowing their trumpets and shouting. The Israelites ransacked and burned the city, sparing only the family of the prostitute Rahab who had helped them. When all was done, Joshua uttered a curse on 'anyone who tries to build this city, Jericho!' (Joshua 6: 26).

Nonetheless, later books in the Hebrew Bible concede that settlement in Jericho persisted, though it was only through divine intervention that the prosperity of the oasis was restored. After the prophet Elijah was taken to heaven, his disciple Elisha picked up his mantle, parted the waters of the Jordan River once again and returned to Jericho, where he remained for a while. The people of Jericho came to Elisha and complained that, although their city was well placed, its waters were bad and the land around it was unproductive. It is possible that the biblical author harks back to the curse that Joshua had placed on Jericho after destroying it several generations earlier. In any case, Elisha agreed to solve the problem. He asked the people of Jericho to bring him a bowl of salt, which they did. He then threw the salt into the spring, citing the Lord as saying 'I have healed this water. Never again will it cause death or make the land unproductive' (2 Kings 2:21). The author concludes the story by saying that the water of Jericho 'has remained pure to this day'.

The location of Jericho was not shrouded in mystery in the same way as Sodom or the grave of Moses were for subsequent pilgrims and explorers. As we shall see, the Jericho oasis was inhabited and even prospered in the Roman era, though the centre of the settlement was no longer Tell es-Sultan, with its thousands of years of history. Writing in the early fourth century CE, the bishop and historian Eusebius of Caesarea (c.260–339) pointed out that the ruins of ancient Jericho were still visible. The oasis remained a site of considerable wealth in the Early Islamic period, when it became known as Ariha, 'fragrant' in Arabic. However, by the nineteenth century, all remnants of prosperity were long gone. Visitors from abroad seldom had kind words for the small village of Ariha, which was situated a few kilometees south of the Tell, or for the ancient remains in the vicinity. The American biblical scholar Edward Robinson, for instance, who visited the site in 1838, noted that it bore little resemblance to the renowned ancient 'City of Palm-trees'.[22] The famous American author Mark Twain, whose account of his pilgrimage to the Holy Land in 1867 reads like a litany of complaints, felt that 'Ancient Jericho is not very picturesque as a ruin. When Joshua marched around it seven times, some three thousand years ago, and blew it down with his trumpet, he did the work so well and so completely that he hardly left enough of the city to cast a shadow.' Twain longed to 'cross the Jordan where the Israelites crossed it when they entered Canaan', but found the river underwhelming: 'many streets in America are double as wide as the Jordan'.[23]

A SITE OF GOD

A year later the British Royal Engineer Captain Charles Warren of the PEF made an early attempt to understand whether the mounds (Tells) of the southern Jordan Valley were artificial and, if they were, what was their composition. Warren did not yet have an understanding of stratigraphy, the archaeological method of discerning successive layers of occupation on the same site that was developed from the 1890s onwards. He had his Arab workmen cut an 8-foot trench across Tell es-Sultan from east to west, mainly so they would not be exposed to the scorching sun. He then instructed the workers to dig down further in a number of shafts. Warren noted that 'very little was found except pottery jars and stone mortars for grinding corn'. He concluded that 'these mounds are formed by the gradual crumbling away of great towers or castles of sunburnt brick'. The PEF's *Survey of Western Palestine* that was published in the early 1880s placed the 'Jericho of Joshua' at Ain es-Sultan and concluded that 'the earliest city must have stood on the Tell'.[24]

John Garstang, who excavated Tell es-Sultan in the 1930s, explicitly sought to demonstrate the reliability of the biblical account of the destruction of Jericho. He interpreted the signs of devastation of what is now understood to be the Early Bronze Age III city and its walls as the result of an earthquake, which aided Joshua's conquest. However, as archaeologist Lorenzo Nigro pointed out much more recently, 'the ruins at Tell es-Sultan are far older than the alleged date of Joshua's conquest'.[25] The Permanent Delegation of Palestine to UNESCO, which submitted an application to recognise Tell es-Sultan as a World Heritage Site in October 2020, presented a similar interpretation: 'Jericho's mighty fortifications had collapsed around 1550 BC, at the end of the Middle Bronze age, before the time assumed for Joshua's conquest.' The Palestinians sought recognition for Jericho not because of its association with the Israelite conquest that Joshua led, but as one of the oldest cities on the planet and one with a long and rich cultural heritage, as attested by twenty-three archaeological layers of ancient civilisations. The nomination of Tell es-Sultan also noted the association often made between the story of Elisha purifying a spring in Jericho and Ain es-Sultan, which emerges beneath the Tell. Elisha's Spring, as it has often been referred to in historical texts, became one more hallowed site on the itinerary of pilgrims and tourists who visited the area over the centuries. As the UNESCO nomination points out, the belief that Elisha brought to an end the infertility of the land and the area's women 'has been perpetuated in the oral traditions of Jericho until today'.[26]

Debates about the historical accuracy of the Bible are not restricted to the vicinity of Jericho. For instance, questions have been raised about the conquests of King David around the Dead Sea. At the height of his power, King David is said to have struck down 'eighteen thousand Edomites in the Valley of Salt', after which he 'put garrisons throughout Edom, and all the Edomites became subject to David' (2 Samuel 8:13–14). Biblical scholars believe the Valley of Salt refers to the shallow, southern basin of the Dead Sea, which in periods of low precipitation, when the level of the lake dropped, could dry up entirely, leaving only a large salt plain.[27]

The minimalist school of thought is cautious of using the Hebrew Bible as a reliable historical source of information. Minimalist scholars argue that large parts of the biblical text were composed long after the events they describe, and that the stories served didactic purposes as well as elite propaganda agendas. The archaeologist Israel Finkelstein, one of the leading proponents of the minimalist school, does not trust the biblical story about the extent of David's geographically expansive kingdom in the tenth century BCE, for which he finds little external corroborating evidence. Finkelstein argues that the story of David's conquests in Edom, east and south of the Dead Sea, was written against a background of events that took place much later, in the early eighth century BCE. According to this interpretation, the earlier

A SITE OF GOD

story about the victory of David gave legitimacy to King Amaziah of Judah (c.798–769). Amaziah is said to have 'restored the boundaries of Israel' to the Lake of the Aravah, one of the biblical names of the Dead Sea (2 Kings 14:25). The Book of Chronicles, which adds more detail about the reign of Amaziah, echoes the story of David, saying that the eighth-century King of Judah similarly struck down 10,000 of his enemies in the Valley of Salt.[28]

Another biblical episode of conflict near the Dead Sea unexpectedly received partial corroboration in the second half of the nineteenth century CE. In August 1868 Frederick Augustus Klein, a native of Alsace who resided in Jerusalem and worked for an Anglican missionary society, took a trip to the region east of the Dead Sea, to the land associated with the Moabites of biblical times. Near the ruins of ancient Dibon (modern-day Dhiban), some 20 kilometres from the lake, he and his guide reached an encampment of the local Bani Hamida Bedouin tribe. To impress their visitor, the hosts took Klein to see a local wonder: a metre-long basalt stone with rounded upper corners. One side of it was smooth but the other had an inscription that consisted of thirty-four lines of text. Unfortunately, the inscription was in a language that Klein was unable to read. He made a sketch of the stone and copied a few of the letters.

Upon his return to Jerusalem, Klein consulted Julius Heinrich Petermann, the Prussian Consul in Jerusalem, who was also a biblical scholar. Judging from the sketch, Petermann believed that the text was written in Phoenician. He therefore made several attempts to purchase the stele while the Bani Hamida who guarded it kept raising the price. News of the discovery leaked to other interested parties in Jerusalem: the French scholar Charles Clermont-Ganneau and the British engineer Charles Warren, who was excavating in the city for the PEF. Clermont-Ganneau hired a local Christian Arab icon painter named Salim al-Qari to go to Dhiban and make a copy of the text using squeeze paper. The story of the partly torn squeeze that Clermont-Ganneau obtained in November 1869 is likely exaggerated. It involves a fight with or among the Bedouins and the squeeze being torn off the stele before it had a chance to dry. The involvement in the endeavour of Salim al-Qari, who was later implicated in the forgery of antiques, raised suspicion among some critics – even today – about the authenticity of the inscription.[29] But there was yet another twist in the plot. The Prussians asked the Ottoman authorities to intervene with the Bedouins and ensure the delivery of the stele. This was happening at a time when the Ottomans were trying to assert their control over Bedouin tribes in Transjordan, resorting to a military expedition in 1867. The friction and mistrust probably accounts for what happened next: before Ottoman troops reached them, the Bani Hamida placed the stele over a bonfire and, when it was sufficiently hot, poured cold water on it so that it shattered into several pieces, which they divided and hid. The Prussians gave up on the whole affair, while Clermont-Ganneau and Warren each purchased a few fragments. The majority of the stele was pieced together, with the remainder of the text reconstructed using Clermont-Ganneau's squeeze.[30]

Once it was deciphered, the stele turned out to be an extraordinary source of information for events that had taken place in the ninth century BCE. It tells the story of King Mesha of Moab, who rebelled against the Israelite kings of the House of Omri who had subjugated the Moabites. The inscription celebrates Mesha's victories over the Israelites and describes the settlements he succeeded in conquering. Many of the place names it mentions – such as Madaba, Nebo and Dibon – correspond to modern-day places east of the Dead Sea, in the Kingdom of Jordan. The stele also tells how Mesha built a bridge over the Arnon, commonly believed to be present-day Wadi Mujib, the river that flows from the mountains in the east through a stunning gorge into the Dead Sea.

A SITE OF GOD

The Moabite Stone, or the Mesha Stele, as it is now known, sheds new light on a battle mentioned in the Hebrew Bible. In 2 Kings 3, the Israelite King Joram – Omri's grandson – joined forces with Jehoshaphat, King of Judah, and the King of Edom to put down a Moabite rebellion led by Mesha. When the battle started, the armies of the three kings had the upper hand. Then, in his desperation, Mesha sacrificed his firstborn son, causing the tide to turn. The biblical author ends the account of the battle abruptly by saying that the Israelites withdrew and returned to their own land. The correlation with the biblical story and the reigns of Joram and Jehoshaphat led modern-day scholars to date the Mesha Stele to 840 BCE or soon thereafter. Most experts believe that the Moabite Stone is authentic, and today it can be found at the Musée du Louvre in Paris.

Those arguing for the authenticity of the Moabite Stone received further support from a 2010 excavation carried out at Khirbet Ataruz. The site, some 14 kilometres north-west of modern Dhiban, is believed to be where Ataroth stood – a settlement that is mentioned twice in the Hebrew Bible. Ataroth was also one of the cities conquered by Mesha according to the Moabite Stone. The excavations revealed remains dated to the ninth and eighth centuries BCE. Among these was a cylindrical stone altar with seven lines of inscribed text written in the same Moabite language as the Mesha Stele. The inscription seems to commemorate the conquest of the city, giving further credence to a lasting Moabite victory over the Israelites.[31]

The Dead Sea played a hidden role in an event that rocked the southern Levant a few generations after the time of Mesha. A number of biblical prophets allude to a massive earthquake that struck Judah and caused considerable damage. When the prophet Amos, a herdsman from Tekoa on the hills west of the Dead Sea, is first introduced, his sermon is said to have taken place two years before the earthquake, while Uzziah was King of Judah and

Jeroboam was King of Israel. The prophet Zechariah also mentions the earthquake in the days of Uzziah, and Isaiah speaks of how the doorposts and thresholds of the temple in Jerusalem shook and how it filled with smoke in the year that King Uzziah died.

The long reign of Uzziah was dated by the renowned biblical scholar and archaeologist William F. Albright to the eighth century BCE. Subsequent scholars believe that the 'Amos earthquake' occurred in or around 759 BCE. Over the last few decades, archaeologists have found evidence of destruction in various Iron Age sites in modern-day Israel and Jordan, which they attributed to a mid-eighth-century earthquake: collapsed walls at Tell es-Safi (biblical Gath), destruction at stratum VI in Hazor and damage to relevant structures at Megiddo and Deir Alla. There were no archaeological traces of damage caused by a biblical-era earthquake in Jerusalem until Joe Uziel and his team uncovered a collapsed wall and a row of smashed vessels underneath it in excavations at the City of David in 2021.[32] But can we be confident that all these geographically disperse findings were caused by the same seismic event?

The Dead Sea helps to shed more light on the matter. Earthquakes can disrupt the layering of sediments on seabeds. In recent years, the retreating shoreline of the Dead Sea exposed outcrops which researchers can access to study anomalies in the lake's sedimentary record. Drawing on data from three coring sites – Ein Feshkha, Ein Gedi and Ze'elim Creek, all on the western shores of the Dead Sea – geologist Amotz Agnon and his team found evidence of not one but two major seismic events in the mid-eighth century BCE. Two earthquakes, one of them with an estimated magnitude of 8.2, or a series of tremors, likely originating from the Jordan Valley, could account for the geographically broad area that was affected, according to the archaeological record.[33]

Ein Gedi, the spring of the kid goat, is mentioned several times in the Hebrew Bible. From these we know that, in addition to offering a source of water and a place where, later on, agriculture

A SITE OF GOD

flourished, there were caves nearby and a wilderness all around. According to the first Book of Samuel, when David was still a young fugitive, he hid from King Saul in a cave near Ein Gedi. Later in the Hebrew Bible, Song of Solomon paints an especially romantic image of the site: 'My beloved is to me a cluster of henna blossoms from the vineyards of En Gedi' (Song of Solomon 1:14). Though these beautiful verses are attributed to the legendary King Solomon, they probably date to a later period, by which time agriculture in the area had taken off. In one place, the Bible indicates that Ein Gedi was also known as Hazazon-tamar (the Dividing or Pruning of the Palm), a possible reference to its date palms, but otherwise the place name is used consistently.

The Book of Ezekiel, which can be dated securely to after the destruction of the First Temple in Jerusalem by the Babylonians in 586 BCE, mentions Ein Gedi as part of its vision for the post-restoration future. In Ezekiel's vision, water flows from the restored Temple, creating a great river that flows eastwards from Jerusalem, all the way down to the Dead Sea. The fresh water from the Temple brings life to the lifeless lake so that it becomes full of fish, and fishermen are able to cast their nets from Ein Gedi. The vision of Ezekiel gives us a hint about an economic activity around the Dead Sea in the Iron Age. The prophet says that, even after the waters of the lake become fresh, the marshes around it will remain salty. Salt was an important commodity in the ancient world. It is quite possible that salt was extracted from evaporation pools around the Dead Sea in the seventh and six centuries BCE, as it was in later centuries.

The earliest archaeological remains at Ein Gedi date to the seventh century BCE. From Josephus and other Roman-era sources we know for certain that there was a prosperous village called Ein Gedi in the oasis where a number of springs and streams converge. The Arabic name of the place, Ain Jidy, made it easy for nineteenth-century European and American explorers to identify it with the biblical site. One of these explorers, the British clergyman

Henry Baker Tristram, described it as 'the most lovely spot on the borders of the Salt Sea', though he found no trace of the vines or the palm trees mentioned in the scriptures.[34] Kibbutz Ein Gedi, established in the 1950s, today stands on a plateau that overlooks the oasis from the south. A quick search on YouTube will bring up enthusiastic tourist guides who will gladly show both virtual and actual visitors where David and his men might have hidden in what is now a nature reserve near contemporary Ein Gedi.

The Dead Sea area also provides the setting for a number of stories in the New Testament. These are all clustered around the shores of the northern half of the lake. John the Baptist received the word of God while in the wilderness, after which he went 'into all the country around the Jordan, preaching a baptism of repentance' (Luke 3:3). At some point after he had baptised Jesus, John was arrested for criticising the local ruler Herod Antipas, a vassal of the Roman Emperor Tiberius. The Gospel of Mark recounts the story of how John was executed during Herod's birthday celebration, at the request of his stepdaughter, but does not say where this took place. According to the first-century-CE historian Josephus, the execution took place at Machaerus, a hilltop palace and desert fortress overlooking the Dead Sea from the east, to which we will return later on. The beheading of John the Baptist is commemorated in paintings by Andrea Solario, Caravaggio and other artists, as well as in the play *Salome* by Oscar Wilde.

After his baptism, Jesus also went into the wilderness like John did before him. He spent forty days there, during which the Devil tried to tempt him. On an unnamed 'high mountain', the Devil offered Jesus 'all the kingdoms of the world' (Luke 4:5). There was a late-Roman Christian tradition associating this story with a mountain a few kilometres west of Jericho. Named after the forty-day period, it is known in Latin as Quarantena or Quarentanae, and in Arabic as Jabal al-Quruntul. A monastery was first established around a cave in the mountain in 340 CE, and was later rebuilt and enlarged during the Crusader period. The Greek Orthodox Monastery of the Temptation, which is still active today, was constructed in the late nineteenth century.[35] There is even a cable car that brings tourists from Tell es-Sultan in Jericho up to the top of the commanding hill.

The gospels tell of Jesus passing through Jericho on his way to Jerusalem. Zacchaeus, a local wealthy tax collector, wanted to see him, but being short in stature and with the view blocked by the crowd, he climbed up a fig-sycamore tree. Jesus told Zacchaeus to come down and astonished the crowd by saying he would spend the night at the tax collector's house, even though the locals considered the latter to be a sinner. Present-day visitors can find a very old tree standing in the garden outside the museum of the Russian Orthodox Church in Jericho. According to local tradition, this is the very same tree that Zacchaeus climbed up.

In sum, the area around the Dead Sea is dotted with sites that bear religious significance and are associated with biblical events, miracles, battles and upheavals. Despite its pilgrimage sites and monasteries, the lake's surroundings also retained an eerie image. The Mishnah, the first major work of Jewish rabbinic literature compiled in the late second and early third century CE, treats the Dead Sea as a place where impure objects, such as kitchenware bearing the shape of the sun or a dragon, could be cast away.[36] As we shall see, for medieval Christian and Muslim visitors the Dead Sea area remained cursed because of its association with the destruction of Sodom and Gomorrah. Then, in the twentieth century, the caves near the Dead Sea added previously unknown pages to our understanding of ancient religious history.

THE DESERT'S SECRET

Clues about the tremendous treasures that the desert around the lake held began to emerge long before the Dead Sea Scrolls became an international sensation. In the third century CE, Origen of Alexandria, an early Christian scholar and theologian, produced the *Hexapla*, a critical version of the Hebrew Bible, which contained the original Hebrew text as well as different Greek translations of it. Only fragments of this compilation survive to the present day, but notable figures in the early history of the church had access to it and held Origen in very high regard. Epiphanius of Salamis (315–403 CE) says that one of the versions of the Bible that Origen used for the *Hexapla* was found 'hidden in wine jars in Jericho with other Hebrew books' at the time of the Roman Emperor Caracalla (198–217).[37]

Later, in the early ninth century, Timothy I of Baghdad, a Patriarch of the East Syrian Church, gave an account of yet another discovery of ancient writings in a cave near the Dead Sea. His description, written shortly before he died in 823 CE, seems eerily similar to the famous discovery in the 1940s, even though the two events were separated by more than a thousand years:

> We have learnt from certain Jews who are worthy of credence ... that ten years ago some books were discovered in the vicinity of Jericho, in a cave-dwelling in the mountain. They say that the dog of an Arab who was hunting game went into a cleft after an animal and did not come out; his owner then went in after him and found a chamber inside the mountain containing many books.[38]

Timothy speculated that these scriptures, which included the Bible and other books in Hebrew script, were deposited by the biblical prophet Jeremiah or one of his contemporaries before the Babylonian conquest of Judah and the destruction of the First Temple in the sixth century BCE.

The next precursor of the Dead Sea Scrolls (as far as we know; there may have been others in between) comes from the nineteenth century and is especially controversial. Moses Shapira was an Eastern European Jew who settled in Jerusalem and converted to Christianity. Among other things, he became a shopkeeper who collected and sold antiques. Alongside Salim al-Qari, Shapira came into disrepute when supposedly Moabite pottery figurines that he had sold to the Prussian government were revealed by the French scholar Clermont-Ganneau to be inauthentic. Shapira then dedicated a number of years to collecting ancient manuscripts. He travelled to Egypt and Yemen and managed to sell hundreds of manuscripts to a number of institutions, including the British Library.

In 1883 Shapira attempted to sell various strips of sheepskin on which a text written in ancient Hebrew letters could still be read. He claimed that he had acquired these strips in 1878 and that they had been discovered by Bedouins 'in caves high up in a rock facing the Moujib (the river Arnon)'.[39] This site was not far from where the Moabite Stone had been discovered ten years earlier. Once he put the strips together and deciphered the text, Shapira noticed that the manuscript was a version of the biblical Book of Deuteronomy, but with a number of significant variations to the standard text. The Jerusalem-based dealer first tried to sell the manuscript in Germany, but several of the scholars who consulted it pronounced it a forgery. He later travelled to London in the hope of selling the manuscript to the British Library. Once again, several experts consulted the manuscript and declared it to be inauthentic. Claude Conder of the PEF sent a letter to *The Times* explaining his reasoning for doing so, as did Clermont-Ganneau, who had deciphered the Moabite Stone and was also involved in revealing Shapira's previous forgeries. In the following year, 1884, Shapira took his own life. Today, the whereabouts of the manuscript, if indeed it survived, remain unknown. This episode was largely forgotten until the discovery of the Dead Sea Scrolls in the late 1940s.

Was the Shapira manuscript an authentic scroll that was composed and hidden in antiquity, something his contemporaries had never seen and were not familiar with? This is an argument that has been put forward by a number of researchers in recent years. For instance, Idan Dershowitz uses the notes left behind by Shapira himself, now kept in the Staatsbibliothek in Berlin, to show how the nineteenth-century dealer was trying to figure out the order in which the strips needed to be placed and the meaning of various parts of the text, therefore indicating he could not have been the forger. Furthermore, by using Shapira's notes, Dershowitz reconstructs the manuscript and posits that it was not only ancient but also an ancestor of the biblical Book of Deuteronomy we are familiar with today.[40]

The debate surrounding Shapira's scroll is likely to continue in the years to come. One issue that those who believe in the authenticity of this now-lost scroll need to account for is where exactly it was found. In his letter to *The Times* in August 1883, Conder made a very good observation, pointing out that a scroll made of sheepskin would be unlikely to survive in good condition for several centuries in a damp cave.[41] As we shall see, the famous Dead Sea Scrolls were hidden in caves near Qumran on the western side of the lake at an altitude of around 330 metres below sea level: an exceptionally dry place where organic matter was more likely to survive. It is not impossible that the manuscript collector Shapira was able to get hold of an authentic ancient scroll originating from the caves in the vicinity of Jericho; as we have seen, such scrolls appear to have been discovered previously in the third and ninth century CE. However, even if that were the case, Shapira certainly lied about its provenance, possibly hoping to benefit from the fame that the Moabite Stone lent to the hill country east of the Dead Sea. No other scrolls have been found in that area, east of the lake.[42]

In the decades following the discovery of the Moabite Stone in the late 1860s, the land associated with the Bible did not turn out

A SITE OF GOD

monumental archaeological finds such as those discovered along the Nile Valley or in Mesopotamia. There was no equivalent to the grandeur of the gates of Babylon that were dismantled, sent to Berlin in crates and carefully reconstructed in the Pergamonmuseum, or to the Egyptian artefacts now held in the Louvre in Paris. Instead, possibly the most important ancient findings in the country came from simple-looking jars tucked away inside caves near the Dead Sea.

There are different versions of the story of how exactly the Dead Sea Scrolls were discovered in late 1946 or early 1947. According to one of the most often repeated, Bedouin shepherds were tending to their flock near Khirbet Qumran by the northwestern edge of the Dead Sea, some 14 kilometres south of Jericho. One of them threw a rock into a cave in the rockface and heard something break. It was getting late, and the shepherds decided to return the next day to investigate. However, one of them, Muhammed edh-Dhib ('the wolf'), was eager to get into the cave first in case there was any gold to be found. Early in the morning he climbed up into what would become known as Cave 1 and found several jars, some with lids, lining the wall. Instead of gold, many of the jars contained manuscripts. The Bedouins took some of the scrolls and brought them to an antiques dealer in Bethlehem. Eventually, they reached the biblical scholar and Hebrew University Professor Eliezer Sukenik, who authenticated them.[43]

Archaeological excavations at Qumran and the surrounding caves began in earnest in the early 1950s. Roland de Vaux, a French Dominican priest and the director of the École Biblique of Jerusalem, led the excavations and was in charge for many years of the official publication of the scrolls in volumes titled *Discoveries in the Judaean Desert*. Over time, some 900 scrolls have been found in 11 caves around Qumran, although almost all of them had disintegrated into thousands of fragments that had to be pieced together, where possible. The scrolls were mostly written in Hebrew, often on sheepskin, and date back to between the late fourth century BCE and the first century CE. A small number were written in Aramaic, and one cave also yielded texts in Greek. Occasionally, papyrus was used rather than parchment. The value of the scrolls for the study of the early composition of the Hebrew Bible cannot be overstated. The scrolls include some 220 copies of the various books of the Bible. They precede the Masoretic Text, the standard version of the Hebrew Bible that dates back to the medieval period and which Jews still use today, by approximately 1,000 years.

While surviving texts of the Hebrew Bible from the medieval period onwards are nearly identical, the scrolls demonstrate that, during what the Jews call the Second Temple period, different versions of the biblical books circulated in parallel. Occasionally, the scrolls vary from the Masoretic Text (or rather, the other way around) in illuminating ways. In the case of the story of Nahash the Ammonite, who besieged the Israelite town of Jabesh-Gilead and was eventually defeated by Saul (1 Samuel 11), one of the scrolls offers more background information than the Masoretic Text, making the biblical story easier to follow. The scroll, a version of the Book of Samuel found in Qumran Cave 4, and therefore given the reference 4QSam, includes a few sentences that must have been omitted at some point as the Bible was being copied. These sentences explain that Nahash was the king of the Ammonites and that he oppressed the tribes of Gad and Reuben, regularly gouging out the right eye of his enemies, until not one Israelite was left east of the Jordan River who had not suffered that fate. This context gives more meaning to Saul's victory over Nahash and helps explain why his leadership became more widely established prior to his being anointed the first king of Israel. The New Revised Standard Version of the Bible, a translation that was published in 1989, includes the sentences recovered from 4QSam as part of 1 Samuel 10.

Some of the scrolls contain non-biblical texts. These texts remain frustratingly silent about the Dead Sea area itself. One imagines the scribes labouring away on the manuscripts while turning their backs to the lake. Some scholars found faint geographic clues in the Damascus Document, of which ten copies were preserved in the Qumran caves.[44] The Document tells the history of an exiled community founded by dispossessed priests of the prestigious Zadokite lineage. It also outlines rules and regulations by which its members ought to live. The 'land of Damascus' mentioned in the Document, Géza Vermès and others argue, should not be taken in a literal sense as referring to the actual Damascus in Syria. Instead, it should be understood as a symbolic place of exile, in the same way as the biblical prophet Amos did. Hence, this place of exile could be understood to mean the region of Qumran itself.[45]

Many scholars believe that the community that lived in Qumran was part of the Essenes, one of the three main Jewish movements around the turn of the first millennium CE. Josephus describes some of their practices, including a preoccupation with hygiene and ritual purification. The Roman author and naturalist Pliny the Elder says there was an Essene settlement near the western shore of the Dead Sea. However, the Dead Sea Scrolls do not mention the Essenes by name. Instead, the scroll known as the Community Rule, of which twelve copies were preserved, uses the term *yahad* ('together').

The first modern scholars who inspected the ruins of Qumran in the late nineteenth and early twentieth century focused on its vast cemetery. The German orientalist Gustaf Dalman, who visited the site shortly before the First World War, remarked that it was 'unusually well suited for a fortress'.[46] The discovery of the scrolls paved the way for different interpretations. For instance, excavations revealed a room that is believed to be a scriptorium because of the remains of inkwells among the debris. Archaeological evidence suggests that the settlement at Qumran was established around 100 BCE. The site offered limited lodgings and did not look as though it could accommodate a great number of people. The size of the communal dining room suggests a capacity of some 100 to 150 members. However, the site also yielded thousands of pottery vessels and extensive tableware, including several hundred undecorated deep bowls. Furthermore, the number and size of the ritual baths was unusually large. Perhaps, de Vaux speculated, the site was home to a larger community with some sleeping outside the settlement in tents or huts? In 2021, Daniel Vainstub put forward a convincing theory for the abundance of public buildings in Qumran despite the site not having lodgings to house a large community. He interprets the Community Rule and Damascus Document scrolls as calling on all members of the *yahad* sect from across the country to congregate at Qumran for the community's annual pilgrimage during the festival of Shavout. Such an interpretation could explain not only the unique archaeological findings at Qumran but also why so many scrolls were held there.[47]

What was the provenance of the scrolls? Were they copied in Qumran or brought to it for protection from libraries in Jerusalem, as scholars like Norman Golb argued? An interdisciplinary study published in 2020 makes an important contribution to the debate. Sequencing of ancient DNA samples that were extracted from scrolls made of animal skin suggests that, while many of the texts were copied at the secluded site of Qumran, others were brought from the outside. For instance, one scroll, a copy of the Book of Jeremiah found in Cave 4 (4Q70), turned out to be made of cow skin. It was clearly brought to Qumran from another part of the country, first because the area around the Dead Sea is not well-suited for cow husbandry, and second because it dates to the late third century BCE, antedating the establishment of the settlement.[48] There is little doubt that the scrolls were hidden in the caves of Qumran by the community itself. The same type of pottery used in Qumran, including cylindrical jars that were unique to it, was found in the scroll caves. Furthermore, as archaeologist Jodi Magness points out, 'some of the scroll caves are in the marl terrace on which the settlement sits and could be accessed

A SITE OF GOD

only by walking through the site'.[49] The scrolls are believed to have been hidden around 68 CE during the Roman suppression of the Jewish Revolt, to which we will return in Chapter 3.

Today, the vast majority of scrolls are held in Israel. Some of these are displayed in the specially built Shrine of the Book at the Israel Museum in Jerusalem. Israel began purchasing Dead Sea Scrolls in the 1950s and got hold of many others in 1967, when the conquest of East Jerusalem brought the large collection of scrolls at the Palestine Archaeological Museum (later renamed the Rockefeller Museum) under Israeli control. A unique finding from Cave 3 in Qumran, written not on parchment or papyrus but on metal, is displayed in the Jordan Museum in Amman: the Copper Scroll, a list of places where various treasures were hidden. Fragments of the Dead Sea Scrolls are also held by various museums and libraries in the United States and in Europe. In March 2020 the scrolls made international headlines once again: a report by a panel of experts concluded that sixteen fragments, purportedly found in the Dead Sea area and purchased for the recently established Museum of the Bible in Washington, turned out to be forged.[50] The Dead Sea Scrolls, it seems, continue to elicit excitement, and retain an aura of mystery.

Technological innovation and new research methods in recent years have paved the way for new revelations in the study of the scrolls, a trend that is likely to develop further in the years to come. For instance, artificial intelligence tools are being used to study minor differences in writing in order to determine how many scribes contributed to a scroll. Advanced imaging technology helps scholars decipher writing on fragments that is too faint to read. Meanwhile, the search in caves for more Dead Sea scrolls has led to other important findings that shed light on the history of the land on both sides of the lake in the Hellenistic and Roman periods, as the next chapter shows.

CHAPTER THREE

❧

A SITE OF WEALTH, WAR AND REFUGE

IN 1969 GIDEON HADAS, an archaeologist and resident of Kibbutz Ein Gedi, west of the Dead Sea, went for a walk by the shore of the lake with his wife. The pair came across a large block of asphalt, large enough to stand or sit on, floating on the water. Although extremely rare in recent decades, blocks of asphalt – or bitumen, as some prefer to call it – that rose from the depths and floated on the surface were common to the Dead Sea in the past. This exceptional natural occurrence, in what the ancient Greeks came to call Lake Asphaltites, was at certain times one of the chief sources of income provided by the Dead Sea. Asphalt was used extensively in the ancient world. It served agricultural purposes, protecting the trunks of date palms and vines from insects. It was also used for veterinary medical purposes, including treating skin diseases of domestic animals, and for insulating boats and vessels containing liquids. Finally, asphalt played a role in the mummification process in Ancient Egypt, especially during the Hellenistic period (323–30 BCE).

The control over precious commodities can often trigger conflict. In the fourth century BCE, the Nabataeans, Arab nomads and traders who had moved into Edom and later went on to build their capital at Petra, harvested asphalt from the lake's surface and sold it across the region. Asphalt proved to be a source of such considerable wealth that, to secure their control over its extraction and trade, the Nabataeans fought a naval battle on the Dead Sea against the successors of Alexander the Great in c.311 BCE.[1]

From the Hellenistic period, which began after the conquests of Alexander, until the late Byzantine period in the seventh century CE, the Dead Sea became a source of wealth and a place where great affluence could be displayed. Alongside the sale of asphalt and salt, the region's agriculture expanded, benefiting from improvements in water-management techniques, large investments by ruling dynasties and access to extensive trade networks.[2] At the same time, the Dead Sea area was also the scene of bitter conflicts as successive kingdoms and empires sought to establish or reassert their control on the lands all around the lake.

THE ASPHALT LAKE

The sources available to us about the Dead Sea area, both written and archaeological, increase gradually as we approach the Roman era. To appreciate the relative textual and material abundance, let us first consider some developments in the preceding centuries. In 539 BCE, Cyrus the Great, the founder of the vast Persian Achaemenid Empire, defeated the Babylonians. The southern Levant came under Persian rule and would remain in their hands until the region's conquest by Alexander the Great in 332 BCE. We know very little from historical sources about the Dead Sea area in this period. Under the Persians, the province west of the lake was known as Yehud. Yehud-stamped jars found in archaeological excavations suggest that Jericho was an agricultural centre. Eleven stamped jars from the Persian period were also found in Ein Gedi.[3]

A SITE OF WEALTH, WAR AND REFUGE

Thanks to more extensive source material, we know more about the story of the Dead Sea during the Hellenistic period. Antigonus Monophthalmus – or Antigonus the One-Eyed – was a prominent commander in Alexander's army. After the great conqueror's death, Antigonus was one of those vying to take control of the vast empire Alexander had created. While extending his territories in the southern Levant, Antigonus sought to prepare boats, take over the harvesting of asphalt from the Dead Sea and use it as a source of revenue for his kingdom, much like the Nabataeans had done. He appointed General Hieronymus of Cardia to oversee this task. However, this arrangement did not suit the Nabataeans. They mustered a force that was reckoned to number 6,000 and, sailing on rafts of reed, attacked the boats sent by Hieronymus with arrows and defeated them. Thereafter, Antigonus gave up on the Asphalt Lake as his attention turned elsewhere.[4]

Hieronymus, the defeated general, was also a historian. Although his works have not survived to the present day, they were used by the first-century-BCE authors Diodorus Siculus and Strabo of Amasia, and it is through them that we can learn how asphalt was harvested from the Dead Sea. Both say that those living around the lake were familiar with tell-tale signs of its impending appearance. Days in advance, a peculiar odour would spread from the lake, and any objects in the area made from silver, gold or brass would lose their characteristic lustre. The asphalt, along with bubbles, would then rise to the surface at irregular intervals. To those who viewed it from a distance, the asphalt block took the appearance of an island. The locals would make their way to the asphalt on rafts, chop it and carry off as much as they could. Diodorus reports that much of the asphalt was sold in Egypt, where it played an important ritualistic and practical role.[5]

The process of mummification involved not only removing the internal organs of the dead body but also drying the corpse and wrapping it with linen and with substances that could assist in its preservation. Two ancient authors, Herodotus in the fifth century BCE and Diodorus four centuries later, give a slightly different list of substances used in the embalming process. Some of these ingredients had to be imported to Egypt from other parts of the ancient Near East. Cedar oil probably came from Lebanon or Cyprus, myrrh from Yemen or East Africa. Cinnamon and cassia must have come from even further afield. Unlike Herodotus, Diodorus specifically mentioned Dead Sea bitumen as one of the ingredients used. The word *mumia* made its way to us through Latin but has its origin in the ancient Persian word for bitumen. However, for many years, scientists questioned whether bitumen was actually used in mummification. Then, in the late 1980s, geochemists Rullkötter and Nissenbaum were able to find traces of bitumen in Egyptian coffins and mummies from different periods, ranging from between 900 BCE and the second century BCE, in the collection of the British Museum. Having analysed their molecular composition, they noticed differences between the earlier and later samples. They were confident that the samples from the Hellenistic period originated from the Dead Sea. Nissenbaum suggests that bitumen was likely used as an external defence against biological decay, worms and bacteria.[6]

Asphalt was not the only thing traded in the Dead Sea area. There seems to have been a sizeable slave trade in the southern Levant already during the late Persian period, with members of different religions being both bought and sold. Much of what we know about this topic comes from a discovery, not far from the Dead Sea, by Bedouins of the Ta'amireh tribe. In 1962 they sold several papyri and other artefacts that they had found in a cave to the Palestine Archaeological Museum in Jerusalem (the sale was made through an antiques dealer from Bethlehem, Khalil Iskander Shahin, better known as 'Kando', to whom we will return later on). The papyri were written in Aramaic, the standard language of

A SITE OF WEALTH, WAR AND REFUGE

scribes in the Persian period, and date to the fourth century BCE. Most of the papyri are concerned with the sale of individual or groups of enslaved people. The circumstances that brought the papyri to where they were discovered are a reminder that the transition from the Persian to the Hellenistic period was not necessarily a smooth one. In 332/331 BCE, shortly after Alexander's conquests, a revolt against the governor he appointed broke out in the city of Samaria in what is today the West Bank. When Alexander's forces set out to suppress the revolt, several people fled from Samaria, taking with them their most valuable possessions, including the above-mentioned legal documents. Based on their names and their place of origin in Samaria, some scholars believe that the documents belonged to Samaritans. However, Jewish and Samaritan names are similar, so it is difficult to differentiate between them. Moreover, in the fourth century BCE, the Samaritans were not yet as distinct from Jews, in terms of their beliefs, as they would become later on. In any case, those fleeing Samaria found refuge in a cave in Wadi Daliyeh, some 14 kilometres north of Jericho. Their hideout appears to have been discovered. Archaeologists believe that a fire lit by Alexander's soldiers at the entrance of the cave may have suffocated those inside.[7]

We also have evidence of enslaved people being traded in the area around the Dead Sea later in the Hellenistic period. A key source on the slave trade comes from the Zenon papyri, a set of administrative records from Ptolemaic Egypt. Zenon was a well-organised bureaucrat who worked for Apollonius, the finance minister of Ptolemy II Philadelphus (284–246 BCE). Zenon kept a detailed record of his business dealings and travel, and, luckily for scholars studying the period, these records were preserved in more than 2,000 papyrus fragments. Zenon must have tucked away his archive safely in Philadelphia, modern-day Faiyum, where villagers discovered it by accident in 1914–1915. In 260 BCE, Zenon was sent from Egypt to the southern Levant to check the books of various landholders. Among other places, he visited Jericho, which was prosperous enough for him to get supplies of grain there. Zenon bought an enslaved girl from the Jewish Tobiad family, and the archives include correspondence about the delivery of more enslaved people by the Tobiads to the elite in Egypt.[8]

The wealthy and politically powerful Tobiad family exercised its influence on both sides of the Dead Sea since the Persian period. The head of the family is likely to have been Tobiah the Ammonite, who is mentioned in the Bible as a rival of Nehemiah, the rebuilder of Jerusalem in the fifth century BCE (Nehemiah 2:10). During the Hellenistic period the descendants of Tobiah mediated between the high priest in Jerusalem and the Ptolemies in Egypt. The Tobiads' estate was situated in Tyros (today Iraq al-Amir) in the beautiful Wadi es-Sir, roughly equidistant from downtown Amman and the Dead Sea. The estate included natural and rock-cut caves, two of which are still fronted with the inscription 'Tobyah' in Hebrew letters. Below the caves stands the impressive Qasr al-Abd ('Castle of the Slave'), an unfinished monumental structure built by Hyrcanus the Tobiad in the late third or early second century BCE. The purpose of this structure, with its columns, enormous blocks of stone, lion carvings and panther-sculptured fountains, has long been debated. Some interpreted it as a temple that may have imitated in some way the Second Temple in Jerusalem. Others argue it was meant to serve as either a fortress, a palace or a mausoleum for the Tobiads.[9] Partly restored first by French experts in the 1980s and subsequently by the Jordanian Ministry of Tourism and Antiques, this little-known gem is well worth visiting.

Large-scale construction in the Dead Sea area developed further under the Hasmoneans. This dynasty emerged out of an armed rebellion against the Seleucid Empire, which had wrested much of the southern Levant from the Ptolemies at the start of

A SITE OF WEALTH, WAR AND REFUGE

the second century BCE. During the nadir of their rebellion, the Hasmonean brothers – also known as the Maccabees – spent a few years in the Judean Desert west of the Dead Sea, where they received some assistance from the Nabataeans. The latter may have taught them a thing or two about the importance of desert strongholds.[10]

In 2022, as part of a survey by the Israel Antiquities Authority, archaeologists found a hoard of coins from this period in one of the caves of Wadi Murabba'at (Nahal Darga in Hebrew), near the north-western shore of the Dead Sea. Inside a cylindrical wooden box was an upper part filled with packed earth and small stones, and beneath it a large piece of purple woollen cloth. At the bottom were fifteen coins minted by King Ptolemy VI of Egypt between 176 and 170 BCE, along with pieces of sheep's wool, probably put there to muffle any noise while the box was being transported. Archaeologist Eitan Klein believes that this treasure was hidden in the wake of the tumultuous events leading up to the revolt of the Maccabees against the Seleucid Empire in 167 BCE. The Book of Maccabees tells of how King Antiochus IV Epiphanes attacked the Jewish Temple in Jerusalem and looted its treasures in 169 BCE (1 Maccabees 1:20–25). The Seleucid king attacked Jerusalem once more the following year, massacring many and destroying the city walls. Many fled Jerusalem and hid in the wilderness (1 Maccabees 2:28–31). The box, preserved for about 2,200 years by the unique conditions near the Dead Sea, may have been left in the cave so that the money could be used once things calmed down, but its owner never returned.[11]

After a lengthy struggle, the Hasmoneans first succeeded in gaining a semi-autonomous status in Judea and later expanded the territories under their control. In a number of campaigns in the late second century BCE the Hasmoneans conquered Perea, the land east of the Jordan River and north-east of the Dead Sea. The Hasmonean rulers, John Hyrcanus I and his son, Alexander

Jannaeus, fought against the Nabataeans, whose kingdom lay south-east of the lake. Under the reign of Alexander Jannaeus (104–76 BCE), a dock surrounded by impressive fortifications was built at Khirbet Mazin, a few kilometres south of Qumran, to assist boats crossing the Dead Sea. In 2002 archaeologist Yizhar Hirschfeld found a staggering 1,735 bronze coins near this site. Dating to the period of Alexander Jannaeus, the coins feature an anchor symbol and inscriptions in both Hebrew and Greek. The site appears to have fallen out of use only a few decades after it was built. Hirschfeld argues that the level of the Dead Sea, which must have been approximately -391 metres at the time of construction to allow boats to dock next to it, probably fell by a few metres, moving the shoreline away and making the docking facility redundant.[12] Today, the reconstructed fortifications of Khirbet Mazin can be seen just east of Route 90, which runs along the western coast of the lake.

One place that thrived under the Hasmoneans was Jericho. The Hasmoneans built a palace complex, not on Tell es-Sultan, but some two kilometres south, near Wadi Qelt. Overlooking Jericho, atop what many Christians later came to associate with Mount of Temptation, the Hasmoneans built the fort of Dok (Duk in some sources), the scene of a dramatic episode of treachery. In 135 BCE, the Hasmonean governor of Jericho, Ptolemy, son of Abubos, hosted the High Priest Simon Thassi and two of his sons for a feast at Dok. Simon was the first leader to have Hasmonean rule recognised by the major powers in the region. He was also the father-in-law of Ptolemy. The Book of Maccabees tells of how, once the guests were drunk, Ptolemy and his men killed them. The historian Josephus adds that, after the murder, Ptolemy tried to seek the backing of the Seleucid King Antiochus VII and to enter Jerusalem. However, he was chased out and besieged at the stronghold of Dagon, likely another name for Dok. Ptolemy endured the siege, probably because the water in the cisterns on

A SITE OF WEALTH, WAR AND REFUGE

the steep slope of the fortress was available only to the defenders, and eventually fled across the Jordan, where the historical record loses track of him. However, the memory of his treachery lingered. Dante Alighieri placed Ptolemy in the *Inferno*, the ninth and deepest layer of hell, in his fourteenth-century *Divine Comedy*. Not much remains of the Hasmonean fortress of Dok today. The peak of the hill was cleared at the beginning of the twentieth century for the building of a church, which was never completed.[13]

In addition to Dok, the Hasmoneans established fortresses at Hyrcania, some 5 kilometres west of Qumran, and on the hilltop of Machaerus, which overlooks the Dead Sea from the east. Josephus says that they also fortified Masada, although save for a few Ptolemaic and Seleucid coins, archaeologists have struggled to find distinct architectural remains from this period.[14] All the fortresses and palaces near the lake were later expanded by one of the greatest builders in the history of the region and the person who took over from the Hasmoneans: Herod the Great.

FROM HEROD TO THE FALL OF MASADA

By the time Herod came of age, the Hasmonean rulers had become clients of Rome. Herod's father, Antipater, was an Idumaean – an Edomite from the region south of Judea whose family had converted to Judaism. Antipater served the nominal Hasmonean ruler but also maintained good relations with Julius Caesar. Herod's mother was Cypros, the daughter of a Nabataean nobleman. As a young man, Herod began his career as the governor of Galilee, farming the taxes for the Roman Senate and cultivating his relations with the Roman general Mark Antony. Then, in 40 BCE, the Hasmonean Kingdom was temporarily overrun by the Parthians, the enemies of Rome, who appointed their own Hasmonean puppet, Antigonus. Herod fled for his life and deposited his family for safekeeping at the fortress atop

THE DEAD SEA

Masada. Antigonus besieged it, but, although those inside were for a while running low on water until the rain filled their reservoirs, he failed to take it. Meanwhile, Herod sought assistance, first from the Nabataeans, and, when it was not forthcoming, he fled through Egypt to Rome. With Roman backing, Herod returned, defeated Antigonus and married the Hasmonean princess Mariamne to legitimise his claim as King of the Jews. By 37 BCE he had captured Jerusalem.

In the first few years of his reign, Herod had to tread lightly. The union between the last Ptolemaic Queen, Cleopatra, and Mark Antony, Herod's patron, meant that any demand made by the two had to be acceded to. Herod therefore lost control over the palm tree groves of Jericho to Cleopatra and had to agree to share with her the profits from the extraction of bitumen from the Dead Sea. Herod's luck changed in 31 BCE as a result of the famous Battle of Actium, where Octavian – soon to become Augustus, the ruler of the Roman Empire – defeated Mark Antony. Ironically, Herod would have fought alongside Mark Antony, had it not been for Cleopatra convincing her lover to send Herod to fight against the Nabataean ruler instead. Cleopatra's suicide following the defeat at Actium meant that, from that point onwards, Herod had to win over and please the master of Rome, Augustus, and his right-hand man, Marcus Agrippa. Herod did so with ruthless and exceptional skill. His rule, which lasted until his death in 4 BCE, was characterised by his execution of several members of his family, on the one hand, and by massive construction projects, on the other.

The vicinity of the Dead Sea was one area into which Herod poured his effort, his money and the talent of his architects. Damage caused by an earthquake in 31 BCE gave further impetus to carrying out construction and renovation works in the area. The Hasmonean palace at Jericho was enlarged and made much more lavish. Possibly in order to impress Marcus Agrippa, who visited the country in 15 BCE, Herod established a complex of three successive palaces, one of which straddled the banks of Wadi Qelt. This complex benefited from a banqueting hall, an enclosed garden and various baths, and it was served by an aqueduct that brought water from springs a few kilometres to the west. Alongside local mudbrick, the palaces were constructed using Roman concrete, a technology that was rarely employed in the southern Levant at the time. The Jericho palaces served the royal family in winter, when the weather in Jerusalem became too cold.[15] Josephus tells us that the gardens of Jericho in this period, watered by Elisha's Spring, were divine and filled with trees that produced fruit of unparalleled quality thanks to the warm climate of the area. To the west, some 250 metres above Jericho and just south of Wadi Qelt, Herod built the luxurious fortress of Cypros, named after his mother. Once again, an elaborate aqueduct was constructed to bring water to this towering site from springs several kilometres away.[16]

East of the Dead Sea, Herod enlarged the fortresses at Machaerus so that it also served as a palace. Pliny the Elder describes Machaerus as 'the most strongly fortified place in Judea', and, indeed, the Hungarian archaeologist Győző Vörös, who excavated the site in the early 2000s, found intact walls that were 8.75 metres high in its western bastion.[17] Nearby, below, on the lake's eastern coast, Herod built a port at Callirrhoe. It was the hot springs of Callirrhoe, reputed for their medicinal qualities, that Herod came to at the end of his days, hoping to treat his many ailments. However, his deteriorating health did not improve. He returned to Jericho, where he died, not before issuing a final malicious order, which was not implemented, to murder various notables upon his death so that the whole country would mourn his parting instead of celebrating it.

Probably the site that benefited most from Herod's attention was the fortress of Masada, a flat hilltop directly overlooking the

Dead Sea from the west. According to Josephus, Herod first fortified Masada in case his ambitious neighbour, Cleopatra, managed to convince Mark Antony to do away with him so that she could take over Judea. Later, once Cleopatra was out of the way, Herod turned Masada into an exquisite retreat. It gained a palatial complex on the triangular northern tip of the plateau as well as a lavish stepped residence and reception hall on the terraces below. Each of the large cisterns constructed atop Masada could hold enough drinking water to sustain a thousand people for one year. The abundance of water meant that Herod's complex was equipped with bath houses and swimming pools. Furthermore, the southern half of the plateau was cultivated with a garden for food crops.[18]

Masada earned its mythical status because of events that took place some seventy years after the death of Herod. Early in the first century CE, what had been the Hasmonean kingdom was formally incorporated into the Roman Empire, becoming the Province of Judea. In 66 CE, tensions around Jerusalem erupted into a full-scale rebellion against Roman rule. The Roman garrison in the city was overrun, and rebel leaders seized various towns and strongholds across the country. One prominent Jewish leader was Yosef ben Matityahu, who would become our main source of information for the First Jewish–Roman War. Appointed as the commander of the Galilee, Yosef surrendered to the Roman general Vespasian after his force was besieged for forty-seven days. Initially imprisoned, he was later freed once Vespasian became Emperor. Because of the Emperor's patronage, he gained the name Flavius Josephus and could dedicate his last few decades to writing and recording the history of the Jews. His first book, *The History of the Jewish War against the Romans*, provides an eyewitness account of the siege of Jerusalem by the troops of Titus, the son of Vespasian, in 70 CE. The book goes on to describe the last stand of a group of extremist Jewish rebels, known as the

A SITE OF WEALTH, WAR AND REFUGE

Sicarii, at Masada and the Roman siege that was placed around it in 73/74, although by this stage Josephus had left Judea for Rome.

Josephus had no love for the Sicarii who, earlier in the revolt, had used their stronghold at Masada to raid and rob the Jews of Ein Gedi while killing those who were unable to run away. After the fall of Jerusalem, some 960 men, women and children barricaded themselves in Masada's fortress. The Romans took their time before attacking this stronghold, capturing Machaerus east of the Dead Sea first. The siege of Masada was commanded by General Lucius Flavius Silva. Silva's force is estimated to have been 8,000 strong. Their first task was to construct an 11-kilometere-long wall all around the bottom of the hill, as well as walled camps for the soldiers. Many of these Roman fortifications are still visible around Masada today. Next, Silva ordered a ramp to be built on the western side of the hill so that his troops could attack the outer walls of the palace-fortress. Eventually, the Romans breached the wall and were preparing for their final assault. Fully aware of the fate that would face them after their defeat, the rebels gathered to discuss what to do next. Josephus narrates two long and harrowing speeches, supposedly delivered by the Sicarii leader Eleazar Ben-Yair, who called on the rebels to take their own lives rather than face suffering and slavery under the Romans. Though some were reluctant at first, Ben-Yair's passionate words swayed them. The men killed their families first. Lots were then drawn, with ten male rebels charged with killing all the other men. Finally, one out of the ten was charged with killing those that remained before committing suicide. According to Josephus, two women hid and did not take their lives. The next morning, when Roman troops broke through the fortifications and were surprised not to meet any resistance, the two women gave them an account of what had happened. While Josephus probably had access to the Roman account of the siege, the words in Ben-Yair's dramatic speech were, in all likelihood, Josephus's. Greek and Roman authors from Thucydides onwards often put what they believed were the appropriate words into the mouths of leaders at speech-worthy points in time.

The works of Josephus, copied and preserved through the medieval period, were well known to the European and North American explorers who toured the Dead Sea area in the nineteenth century. The German explorer Ulrich Jasper Seetzen found Roman-era fortifications at the site the Arabs called Sebbeh in 1806. He identified this site as ancient Masada, as did the American explorers Edward Robinson and Eli Smith in the late 1830s.[19] In the second half of the nineteenth century, a number of European explorers began to record and map the remains of the fortress. In 1927, a major earthquake struck the area, damaging the Roman siege ramp and what remained of a platform at its top. The earthquake alongside erosion account for why, presently, the ramp only reaches some 15 metres below the edge of the cliff.[20]

The largest-scale excavations at the site were carried out by the Israeli general-turned-archaeologist Yigael Yadin between 1963 and 1965. The findings gave credence to the reputation of Herod as a lover of grand architecture and of a lavish lifestyle: the plateau was surrounded by a wall, approximately 1,300 metres long, with 27 towers overlooking the cliffs. The palace complex included elaborate mosaics with floral motifs, showing olives, pomegranates and grapes. In the storerooms the excavators found storage jars that included, in some cases, the remains of seeds, nuts and fruit. Some of the jars bore Greek or Latin inscriptions describing their content and provenance: Italian or Aegean wine, fish sauce from Spain. Even the name of the recipient, 'King Herod of Judea', was sometimes mentioned. As for the period of the revolt against the Romans, Yadin's expedition found coins minted by the rebels. These bore ancient Hebrew letters and recorded the year since the outbreak of the revolt. The rebels also brought to Masada several scrolls. Yadin's excavation uncovered fifteen parchment scrolls, written in the square Hebrew characters that are still in use today, and one papyrus fragment in the ancient Hebrew alphabet. The texts included works that are not considered canonical in Judaism, such as the Book of Jubilees and Ben Sira, and a liturgical composition known as Songs of the Sabbath Sacrifice.[21] The discovery of these texts, copies of which were also found in Qumran, raised the question: did sectarians from the community in Qumran flee and seek refuge in Masada? Recent analysis of DNA samples that were extracted from scrolls from Qumran and Masada respectively suggests that the sheepskin that was used to create the parchment from the two sites came from different groups of sheep.[22] We can therefore assume that the texts that were found in Masada were known and used throughout Judea and were not restricted to the *yahad* community in Qumran alone.

The excavation on Masada also uncovered ballista stones and several signs of violent destruction. Yadin was especially moved by the discovery of three skeletons. One of them was evidently that of a woman, her scalp intact owing to the extreme dryness of the area, and her black, braided hair not betraying the almost 2,000 years that had passed since she died. By her side were Roman-era leather sandals. Following Josephus's account closely, Yadin interpreted twelve ostraca – broken pieces of pottery – with Hebrew names on them, including that of 'Ben Yair', as the lots that determined which of the men would take the lives of their comrades. Other archaeologists were more sceptical about the accuracy of Josephus's account, suggesting the ostraca may have served as tags used in the distribution of food among the rebels. Masada also did not yield quite as many skeletons to accord with the description of the mass suicide. It is possible that the Roman garrison that remained in Masada disposed of the vast majority of the bodies by burning them. Evidence of continued fighting within the fortress, after the external wall was breached, cast further doubts over the mass suicide that Josephus recorded.[23] Regardless of this debate between archaeologists, Masada remains a magnet for tourists.

When tourism in the area peaked in the late 2010s, the site overlooking the Dead Sea attracted almost 800,000 visitors each year.

THE CURIOUS DEAD SEA AND ITS MYSTERIOUS FRUITS

A number of Greek and Roman philosophers, geographers and historians remarked on the unique nature of the Dead Sea. What often attracted their attention were the buoyancy of the lake, the asphalt that emerged from its depths, the strange taste and odour of the water, and the lack of living creatures in it.[24] Aristotle's *Meteorology*, written around 340 BCE, is an early example. Without having visited the area, Aristotle had heard that even if a bound man or beast would be thrown into the lake, they would float on the surface rather than sink. This helped him to conclude that the density of a fluid is greater when a substance is mixed with it.[25] Writing a few centuries later, both Strabo and Pliny the Elder noted the peculiar buoyancy of the Dead Sea, although, much like Aristotle, they had not seen it. Pliny reported that even bulls and camels would float on the lake's surface without sinking.[26] Josephus, who knew the area of the Dead Sea very well, recounts how Vespasian wanted to test the buoyancy of the lake. The Roman general ordered that people, perhaps captives, who did not know how to swim and who had their hands bound be thrown into its depths. The poor human guinea pigs nonetheless floated to the surface as if they were carried upwards by a gust of wind.

Ancient authors were also interested in the flora around the Dead Sea. When writing about biblical Sodom, Josephus says that a remainder of the great, divine fire that destroyed the city could still be found in one of the fruits that grow in the area. The fruit, he says, looks as though it is fit to be eaten. However, if one were to pluck it, the fruit would turn into smoke and ash. This fruit became known as the 'apple of Sodom', which later travellers visiting the lake often sought to identify. The Scottish missionary John Wilson, who visited Jericho in the 1840s, says that, according to local Arab folklore, this plant once bore excellent fruit. However, Lot cursed it, because of the wickedness he had encountered, and from that day on it was doomed to produce bitter fruit. The locals therefore called it 'Leimun Lut' – Lot's lemon.[27] Today, the 'apple of Sodom' is recognised as *Calotropis procera*, a small tree which grows in various parts of Asia and Africa. Rather than ash, the inside of the fruit contains inedible seeds and fibres. The latter may have been used in antiquity as thread wicks to light lamps.

Of much more value, and therefore of far greater interest to Roman-era authors, was the legendary Dead Sea balsam. Pliny, a keen botanist, discusses this plant at length. He describes it as an evergreen small tree or bush with pale leaves. Its great value came from its sap, which was produced by careful incisions in the plant's bark with a sharp tool. Tiny droplets of sap could then be collected. The sap of the balsam was used as a basis for many perfumes. Pliny mentions it as a component in the special 'royal perfume' of the Parthian kings. He also notes balsam was used for medicinal purposes, as do Diodorus, Strabo and other ancient authors. According to the slightly later Jewish Talmud, it was used as an oily ointment.[28]

An aura of mystery still shrouds the balsam today. It was referred to by different names: *balsamum* in Latin, *afarsemon* in Hebrew. In Greek, the sap was called *opobalsamon*. Modern botanists suggest that the Dead Sea balsam belongs to the genus *Commiphora* but are divided with regard to the exact species. In any case, it was a very rare plant. Pliny says that nature had only bestowed it upon the land of Judea and, even there, it was only grown in two gardens: in Jericho and Ein Gedi. Josephus hints that it had been brought to the country from Egypt or Ethiopia by the biblical Queen of Sheba during the reign of King Solomon. If the balsam was indeed *Commiphora gileadensis*, as some modern botanists suggest, then at least the geographic origin in the story that Josephus recounts is correct, as this desert plant does grow in certain places in southern Egypt, Sudan and the Arabian Peninsula. Because of the balsam's great value, the gardens where the plant grew belonged to the kings of Judea and, later on, to the Roman emperors. While she had the ear of Mark Antony, Cleopatra of Egypt temporarily appropriated from Herod the palm groves of Jericho, where balsam was also grown. During the Jewish revolt against the Romans, the rebels tried to destroy the balsam plants, while the Romans sought to protect them. Pliny says that Vespasian and Titus brought the plant to Rome to be carried in their triumphal procession, marking the subjugation of Judea. After the revolt was crushed, balsam cultivation was expanded. We know from archaeological discoveries that a Roman garrison was stationed in Ein Gedi for several decades. The cuttings of the plant, known as *xylobalsamum*, became an article of merchandise as well, though these were the cheapest product of the balsam tree. It seems that, over time, inauthentic items abounded. The physician Galen (128–200 CE) had to travel to Judea to buy the original balsam. By the time of Eusebius, in the fourth century, the plant was also cultivated in Zoar, at the southern end of the Dead Sea.[29]

Despite the rich textual descriptions of the balsam, archaeologists have struggled to find where exactly around the Dead Sea it was grown and produced. A papyrus fragment found on Masada, likely coming from the Roman garrison that remained in the fortress after the revolt, mentions *xylobalsamum*. Potsherds from Masada and other sites have 'balsam' written on them in either Hebrew or Aramaic. A juglet that was uncovered during excavations in Qumran contained an oil that does not resemble any plant oil known today, but it was impossible to determine whether it was the legendary balsam. The remains of the third-century synagogue that was uncovered in Ein Gedi included an inscription in Aramaic that forbids community members to reveal the 'secret of the village'. Some have argued that this secret must have related to the method of growing balsam and manufacturing its perfume, but other archaeologists have cast doubt about this interpretation. Gideon Hadas excavated forty ancient agricultural terraces in the southern side of the Ein Gedi oasis in order to trace which crops were grown there. He found carbonised plant remains of date palms but did not trace remains that could be identified with the balsam.[30]

There are fewer question marks when it comes to date palms, the most widely grown agricultural crop in the Dead Sea area in antiquity. Pliny praises the palm groves of Ein Gedi. Josephus does the same for those in the Jericho oasis. He points out that several types of date were grown in the area, differing from each other in taste and name. When pressed, the best of these yielded an ooze as sweet as honey. Unlike the archaeologically elusive balsam, with date palms we also have plenty of physical evidence to corroborate the descriptions left by ancient authors. As excavations in Jericho, Qumran, Masada and the many caves in the cliffs near the Dead Sea have shown, dates were very widely consumed in the Roman period. Moreover, archaeologists have found several ropes and baskets made entirely from intertwined date-palm fibres. Luckily for the date lovers among us, it seems that one type that was popular in the Dead Sea in antiquity is poised to make a comeback.

THE TIME CAPSULE

'Methuselah' is not very old. Named after the biblical symbol of longevity, the palm tree Methuselah was planted at the Arava Institute for Environmental Studies in Ketura, southern Israel, in 2005. What is special about this palm is the seed from which it was germinated. And the seed is very old indeed. Uncovered during Yadin's excavations on Masada in the 1960s, it originally dates back to the Roman era. In the decades since, hundreds of ancient date seeds were recovered in archaeological excavations near the Dead Sea.

The idea to germinate ancient date seeds came from Dr Sarah Sallon, a physician and ethnobotanical researcher at the Hadassah Hospital in Jerusalem. She was inspired to revive local plant variants that had become extinct after learning of the successful germination of a 1,300-year-old lotus seed which had been found in China. 'Methuselah' was a resounding success, but it had one major flaw. Date palms are dioecious, meaning they are either female or male. Because 'Methuselah' is male, it produces the pollen needed for fertilisation but cannot grow any dates. Therefore, Sallon tried to convince the keepers of the relevant archaeological collections to share more ancient seeds so that a female palm tree could be grown. Eventually, Sallon succeeded. Her project partner, Dr Elaine Solowey, who specialises in desert agriculture and, as I have seen, can grow anything, managed to germinate six more seeds: 'Adam' from Masada; 'Jonah', 'Uriel', 'Boaz' and 'Judith' from Qumran; and, crucially, 'Hannah' from Wadi el-Makkukh, north-west of Jericho. The remarkable durability of the ancient seeds is attributed to some of the unique environmental conditions of the Dead Sea area: low precipitation, very low humidity and, at several hundred metres below sea level, the thickest atmosphere on Earth, with a unique radiation regime.[31]

Having been pollinated by 'Methuselah' and 'Adam', in 2020 'Hannah' produced her first dates. She continues to produce more fruit, and these are carefully picked in the late summer and stored for research purposes at the Arava Institute.[32] Sallon and Solowey hope to have these ancient Judean dates reintroduced on a large-scale basis. 'Date palms that have been germinated from ancient seeds could, in the future, be used to help select new varieties of dates that are more adapted to climate change and pathogen invasions than modern date varieties', says Sallon.[33] I had the honour and the privilege to try one of these wonderful dates. Its taste is similar yet distinct from present-day Medjool dates and is also reminiscent of Barhi dates that are grown in Iraq. One could easily see why Josephus had so much praise for the dates that grew in the Dead Sea area.

In a way, the Dead Sea area acts as a time capsule. It preserved not only organic matter like date seeds but also several written texts, even beyond the famous Dead Sea Scrolls. The ravines, or wadis, that flow into the lake from the west are dotted with caves. These hard-to-reach caves served as a place of refuge in times of rebellion and conflict. Fragmented texts were found in these caves, alongside various other artefacts, such as items of clothing, utensils, coins, weapons and, at times, the skeletons of the people who brought them there. The artefacts, and especially the written texts, help us reconstruct and fill in gaps in the history of Roman rule in the southern Levant. Before the 1950s, the chief sources of information for this period were historians such as Josephus. Thanks to the unique environment of the Dead Sea, we now know more, with new discoveries announced as late as 2023.

Let us first tend to the story behind the recovery of the ancient texts, a cat-and-mouse game between archaeologists and Bedouins. The discovery of the Dead Sea Scrolls, and the growing international interest around them, prompted the Bedouin tribes west of the lake to search the many caves in the area for more ancient written texts. In autumn 1951 Bedouins of the Ta'amireh tribe presented the Palestine Archaeological Museum in East Jerusalem with two fragments of parchments with texts in Hebrew and Greek. Soon afterwards, Father de Vaux, who had already begun to excavate Qumran, was approached by the antiques dealer Khalil Iskander Shahin, or 'Kando', who offered to sell a few other fragments of parchment and papyrus. The latter claimed that these came from Cave 1 near Qumran, but de Vaux was not convinced, because neither their appearance nor their material was similar to the Qumran scrolls. De Vaux was also approached by several members of the Ta'amireh tribe, who each showed him several fragments and various coins. Determined to identify the provenance of all these newly discovered texts, he negotiated with them for several weeks. Finally, in January 1952, the Bedouins agreed to escort him to the place where the texts and artefacts had been found: four difficult-to-reach caves in Wadi Murabba'at, 'one of the least accessible places in Palestine', some 18 kilometres southwest of Qumran as the crow flies. 'When the caves came in view', de Vaux later recalled, 'we were surprised to see coming out of these holes about 34 illegal diggers, who had been working there for several days'.[34] Despite the extensive rummage that had already taken place, an excavation in the caves led by de Vaux uncovered remains ranging from the Chalcolithic to the Roman periods. The latter included several texts that were written in Hebrew, Aramaic, Greek and Latin on papyrus, leather, parchment, potsherds and even paper. They included biblical texts, profane letters, contracts, literary and historical works, alongside administrative and military documents.[35]

Another cave discovery further south by another scroll aficionado would help bring to light the full importance of the texts of Wadi Murabba'at. Yigael Yadin's biography was closely entwined with the Dead Sea Scrolls. It was his father, Eliezer Sukenik, who had purchased the three scrolls that had been found in Cave 1 of Qumran back in 1947. In 1954 Yadin himself arranged for the State of Israel to purchase four additional scrolls. These had been put on sale through an ad in the *Wall Street Journal* by Athanasius Samuel, the Metropolitan of the Syriac Orthodox Church in Jerusalem, who had relocated to the United States. Yadin based his PhD thesis on the War Scroll from Qumran, for which he earned an Israel Prize in Jewish Studies in 1956. As noted above,

A SITE OF WEALTH, WAR AND REFUGE

Yadin discovered several scrolls while excavating Masada between 1963 and 1965. Then, in June 1967, while the Six-Day War was still raging, Yadin used his capacity as the military advisor to Israel's Prime Minister, Levi Eshkol, to get hold of a scroll that, for many years, had remained in private Palestinian hands. He arranged for an Israel Defense Forces (IDF) officer to visit the antiquities dealer 'Kando' in Bethlehem. Under pressure, Kando retrieved and handed over a scroll wrapped in cellophane and kept in a shoebox in a special hiding place. It would become known as the Temple Scroll. Years later, Yadin published the full text of the Temple Scroll in the original Hebrew and in an English translation, along with substantial explanatory material.[36]

However, in terms of changing modern understanding of the past, texts that an expedition led by Yadin had uncovered in a cave near Ein Gedi in 1960–1961 remain in a league of their own. It was part of a large-scale effort in the caves of the Judean Desert, prompted by the discovery of scrolls on the Jordanian side of the frontier, in Wadi Murabba'at, and concerns about the clandestine discovery of scrolls within Israeli territory and their sale by Bedouins. Four archaeological teams were tasked with exploring and excavating caves in six wadis between Ein Gedi and Masada. Although a retired IDF general, Yadin was still considered a junior archaeologist at the time and received the least coveted area: the northern flank on Nahal Hever.

After surveying several nooks and crannies in the area they were assigned, Yadin and his team decided to focus on a large cave in the cliffs of Nahal Hever. The remains of a Roman camp at the top of that cliff, controlling any attempt to enter or leave the cave, suggested it had been of considerable importance. The cave, some 150 metres long, had already been surveyed by previous Israeli archaeologists and had clearly been visited by Bedouins seeking antiques and scrolls as well. Nonetheless, by exploring every hidden corner in the cave, Yadin's expedition was able to find bones and skulls that had been gathered in baskets. These belonged to three men, eight women and six children who must have died in the cave. Sometime after they perished, a family member or an acquaintance must have found them in the cave and tended to the remains, using the baskets in lieu of an ossuary. The cave also yielded Roman-era rugs, tunics and sandals, coins with Hebrew writing that were minted in the first half of the second century CE, arrow heads, keys and another basket with kitchenware made of bronze. Some of the bronze vessels had on them depictions of pagan deities that are strictly forbidden in Judaism. Yadin, who later found out that some of these bronze objects were likely produced in Capua, Italy, assumed that they originally belonged to Roman soldiers and later fell into rebel hands. The Jewish rebels deliberately defaced the deities, thereby rendering the vessels usable. Most important of all, the excavators found two sets of archives that were hidden in a leather waterskin and a leather bag.[37] The papyri in these archives, along with the texts found in Wadi Murabba'at and, subsequently, in thirty other refuge caves across the Judean Desert, completely changed our understanding of the Second Jewish–Roman War, better known as the Bar Kokhba Revolt (132–135 CE).

Before these archaeological discoveries, surprisingly little was known about this revolt. Josephus had died around 100 CE and was no longer around to record it. The Greco-Roman historian Cassius Dio (c.165–c.235 CE) briefly mentions how this revolt was brutally suppressed by Emperor Hadrian, claiming that 580,000 men were slain, while the number of those who perished by famine, disease and fire was beyond counting. Once the revolt was crushed, he added, 'nearly the whole of Judaea was made desolate'.[38] Indeed, under Hadrian, the province of Judea changed its name. Using the place name 'Palestine', which had been employed by Greek and Roman authors in previous centuries, the province became known as Syria Palaestina.

A SITE OF WEALTH, WAR AND REFUGE

What sparked the revolt was unclear. According to Cassius Dio, it was Hadrian's decision to build a Roman colony, Aelia Capitolina, on the ruins of Jerusalem, razed during the First Jewish–Roman War. However, Eusebius in the fourth century CE argued that Aelia Capitolina was only constructed after the second revolt was crushed and as a consequence of it. The leader of the revolt also remained enigmatic. The few remaining ancient Jewish and Christian sources that mention him were clearly divided about his merits and even about his name. In some he appears as Bar Kokhba, which evokes the image of a rising star. In others he is referred to pejoratively as Bar Koziva, which brings to mind a false prophet. Evidence from refuge caves near the Dead Sea helped to resolve some of the mystery.

A refuge cave in Nahal Michmash (Wadi es-Suweinit in Arabic), some 16 kilometres west of Jericho, yielded coins minted in Aelia Capitolina alongside coins minted by the Jewish rebels. Hence, Aelia Capitolina already existed by the time the rebels fled to this cave. We can assume that the founding of a Roman colony – likely coinciding with Hadrian's visit to the country in 130 CE – on the ruins of Jerusalem and its sacred temple angered the Jews greatly and played a part in the outbreak of the revolt.[39] As for the leader of the uprising, both de Vaux in Wadi Murabba'at and Yadin in what came to be known as the 'Cave of Letters' in Nahal Hever found letters dictated by Bar Kokhba to local commanders. His real name was Shimon Bar Kosiba. The letters were written by different hands in Aramaic, Hebrew and Greek. Bar Kosiba refers to himself as Nasi Israel, a title for a leader with messianic overtones that is used in Ezekial's vision in the Bible and which, in modern times, has come to mean 'President'. The letters suggest that Bar Kosiba had taken over the precious balsam groves of Ein Gedi, which had belonged to the emperor, and that he subleased parcels of land in the area.[40] They also show how the rebel leadership tried to assert its independence from the Romans, establishing their own currency with fixed conversion rates and creating their own calendar, taking the beginning of Bar Kosiba's rule as its starting point.

In his letters to the commanders of Ein Gedi, Bar Kosiba comes across as a harsh and impatient leader. He orders them not to give shelter to the people of Tekoa who evaded military service. He also demands that they send to his headquarters various goods, including salt, of which the Dead Sea had plenty. Salt was a precious commodity in the Roman Empire. It was used as a method to pay soldiers their wages – indeed, the word salary comes from *salarium*, a payment made in salt (*sal*). Bar Kosiba often threatened his subordinates with punishment should his orders be ignored. Remote Ein Gedi, it seems, was far removed from where much of the fighting took place and was therefore called upon to supply provisions and hand over fugitives. In a desperate letter, probably written shortly before the end of the rebellion, Bar Kosiba chastises the people of Ein Gedi: 'In comfort you sit, eat and drink from the property of the House of Israel and care nothing for your brothers.'[41] The objects that were found with this collection of letters – items of jewellery, a bottle of perfume and a mirror – suggest that they were brought to the cave by the wife or another woman in the household of one of the commanders of Ein Gedi to whom the letters were addressed.

Another woman who fled to the same cave, bringing with her some of her most prized possessions, was Babatha. It is very rare that we can learn so much about a second-century-CE woman who did not belong to any social or cultural elite. Fortunately, Babatha kept a very organised archive, which consisted of thirty-five papyri of legal and administrative documents. Through these we can not only reconstruct her biography but also get a glimpse of the social history of the area around the Dead Sea during this period. Babatha was born and raised in the town of Mahoza at the southern end of the Dead Sea, near Zoar, where her family had purchased some palm groves. Babatha was Jewish, but the area where she was born was still part of the semi-autonomous Nabataean kingdom. In 106 CE, Emperor Trajan annexed the Nabataean kingdom and turned it into the Roman province of Arabia. Many of the documents in Babatha's archive are in the Nabataean language, while the rest are in Aramaic and Greek. Babatha was married twice and twice widowed. Her second husband was from a land-owning family from Ein Gedi. The documents attest to legal disputes between Babatha and the families of both her husbands. She had been wealthy enough to lend money to her second husband and, after he died, sought to take over his land in Ein Gedi in lieu of the unpaid debt. Despite the legal disputes, Babatha remained in Ein Gedi once the Bar Kokhba Revolt began. When the revolt drew to its fatal end, Babatha, along with a few other residents of Ein Gedi, fled to the cave in Nahal Hever. They were nonetheless discovered by Roman troops who must have been engaged in mopping up any remnants of the rebellion. Instead of trying to mount an assault on the hard-to-reach cave, the Romans set up a garrison in a spot above that controlled the only exit from it. Though they had brought dates, olives, pomegranates and other provisions with them, Babatha and the other fugitives must have died eventually from lack of water or starvation.[42]

There have been more sensational discoveries in the refuge caves west of the Dead Sea in recent years. For instance, in 2023, archaeologist Asaf Gayer and his team returned to one of the caves above Ein Gedi to reexamine a stalactite with an incomplete ink inscription written in ancient Hebrew script that had been discovered fifty years earlier. While in the cave, they spotted what appeared to be a weapon in a narrow crevice. Eventually, four very well-preserved Roman swords with iron blades were extracted. Three of these have been identified as Roman spatha, or long swords, measuring between 60 and 65cm. Archaeologists assume that the swords were booty captured by Jewish rebels and hidden away in the second century CE, possibly during the Bar Kokhba Revolt. Thanks to the Dead Sea time capsule, even the swords' wooden scabbards were preserved.[43]

THE BYZANTINE SPRING

The transition from the Roman to the Byzantine period is not so clear-cut. The Byzantine Empire was, in fact, the Eastern Roman Empire. Under the reign of Diocletian (284–305 CE), the empire was divided into two halves and governed by co-emperors. Though temporarily reunited under his successor, Emperor Constantine I, the Eastern Roman Empire followed a very different and much longer path than its western counterpart. In 330, Constantine transferred the capital of the empire to the site of the ancient Greek colony of Byzantium on the western bank of the Bosporus. Constantine revoked the persecution of Christians ordered by his predecessors and made Christianity one of the empire's permitted religions. He also introduced the *solidus*, a new gold coin that remained in use for centuries. The new path that the Eastern Roman Empire embarked on appears to have coincided with favourable climatic conditions, bringing a period of relative prosperity to the area around the Dead Sea that lasted from the fourth to the early seventh century CE.

During the Byzantine period new settlements were established across Palestine and Transjordan, and the population grew to levels that were not surpassed until the end of the nineteenth century. Even sparsely populated areas such as the Negev Desert in the south-west or the plain east of the Dead Sea, near Karak, saw the establishment of new settlements and the expansion of agriculture. A number of different factors contributed to this growth. First, the growing popularity of Christianity turned hitherto peripheral provinces of the empire into a holy land, a place where one could see the sites where the biblical stories took place. There was certainly

A SITE OF WEALTH, WAR AND REFUGE

a growth in the number of pilgrims visiting the Dead Sea area, alongside construction of large-scale facilities to cater for them. As we have already seen, between the fourth and sixth centuries, churches and monastic complexes were established on Mount of Temptation near Jericho, in Bethany Beyond the Jordan and on Mount Nebo. A sanctuary dedicated to Agios (Saint) Lot was established near Zoar. Judging from the lavish and worldly mosaic at Mount Nebo, which depicts, among other things, a zebra, an ostrich led by a black man, and a lion, as well as hunting scenes, the monastic complex enjoyed a degree of affluence, presumably thanks to either generous or simply very many donations by pilgrims.

Many new monastic communities were established around holy men who, while seeking solitude and an ascetic way of life, nonetheless attracted a following. As Hagith Sivan put it, 'From the fourth century onward the arid hinterland lying between Jerusalem and the Dead Sea swarmed with humans, ascetics seeking solitude, pilgrims seeking ascetics, monks seeking salvation.'[44] In the area between Jericho and the Dead Sea alone, some nineteen monastic cell-clusters, ranging in size from six to forty-four cells, have been identified. These were established between the middle of the fifth and the middle of the sixth century. Byzantine-period anchorites – religious recluses – inhabited some of the caves near Qumran, as well as Qumran itself. There was also a Byzantine settlement a short distance away, by the shores of the lake at Ein Feshkha, with the small oasis possibly serving as a garden for the anchorites. Further south, a monastery was even established on Masada. Often, monastic sites were selected owing to their connection with stories from the Bible. For instance, west of Jericho, on a cliff above Wadi Qelt, monks settled around a cave where the prophet Elijah was believed to have taken refuge and been fed by ravens. Known as the Monastery of Saint George, or the Monastery of Choziba, it still hangs onto the cliff and is active today.[45] Access to water also determined the choice of sites.

There is a correspondence between places where Euphrates poplar trees continue to grow today and sites with Byzantine archaeological remains. One can find poplars in well-watered areas where Byzantine monasteries used to stand in the vicinity of Jericho.[46]

Another reason for the prosperity of the Byzantine period has to do with relative stability and safety. From the second century CE onwards, the Romans built military outposts – forts, fortlets and watchtowers – along the *limes*, the frontier of the empire, east of the Dead Sea and further south all the way down to the Red Sea. The ruins of the legionary fortress of al-Lejjun, north-east of Karak, is one example of these fortifications which can still be seen today. Another is the nearby well-preserved fort at Qasr Bshir. Eusebius says of the area near the Arnon (Wadi Mujib) that 'there are garrisons of soldiers to keep guard from all sides because of the fearful nature of the area'. It seems that the area was fearful, and the fortifications were built, because of the Saracens, a vague name which the late Romans and Byzantines used to refer to Arab tribes. The *limes* should not be taken to mean a fortified line like a modern-day border. The aim of the outposts was not to prevent raids altogether. Indeed, raids could originate from within the territory ostensibly under imperial control. Instead, the forts helped to protect settlements and lines of communication. The outposts served as sanctuaries for settled populations. Manned by mounted troops, they were also able to provide armed escort. The outposts were placed in the vicinity of natural springs, water courses or artificial reservoirs.[47]

Evidence for the impact of climatic conditions on settlement expansion in the Byzantine period took some time to amass. In the late 1980s Rehav Rubin still estimated that the expansion of agriculture resulted from improved water management techniques rather than a wetter climate.[48] While efficient use of runoff water became more widespread than in earlier or later periods, and certainly assisted farming, we are now reasonably confident that

A SITE OF WEALTH, WAR AND REFUGE

the period between 300 and 600 CE also saw higher precipitation. Pollen data from Dead Sea cores confirms that agriculture flourished, for instance with olive cultivation expanding considerably in the Judean Hills. A core from the south-western edge of the lake also attests to increased agricultural activity in its catchment area in the Negev Desert. Lake-level reconstructions, based on radiocarbon ages of organic remains from exposed sedimentary sequences along the Dead Sea shores and other data, vary slightly from one another. However, there is consensus that, after a fall around the first century BCE, the lake filled up, so that during the heyday of Byzantine rule, its level was higher than -390 metres.[49]

Yizhar Hirschfeld was even able to find archaeological evidence of this rise in the level of the Dead Sea. While excavating the Ein Gedi oasis, he found the remains of a Roman bathhouse which was dated to between the two Jewish revolts (70–130 CE). The floor of the bathhouse was at an elevation of -389 metres. Above it was a layer, about 1 metre thick, composed of Dead Sea sediments and alluvium washed down from the hills. In other words, while other parts of the oasis were bustling with agricultural activity during the Byzantine period, the area of the first-century bathhouse had been submerged under the rising level of the lake. Deposits found on the walls of the first-century-BCE fortified dock at Khirbet Mazin also suggest that the Dead Sea had risen after the site fell out of use. Hirschfeld estimates that the lake rose up to at least -387 metres.[50] The rising level of the Dead Sea in the Byzantine period is indicative of a significant increase in rainfall in its large drainage area.

That the Dead Sea area was prosperous during the Byzantine period is apparent from the earliest surviving cartographic record of Palestine and Transjordan: the Madaba map. Madaba is a town that sits in a fertile plain some 30 kilometres south-west of Amman and 20 kilometres east of the Dead Sea coast. In 1890, renovations in Madaba's Greek Orthodox Church of St George revealed the remains of a Byzantine-era structure. These remains included a large but partially damaged mosaic map, which likely sustained further damage during the renovation works. The surviving sections of the map contain 150 inscriptions in Greek giving place names and marking the boundaries between various territories. Most scholars agree that the map dates to the mid-sixth century. The purpose of the map has long been debated: some interpret it as a guide for pilgrims to the Holy Land, possibly based on the works of Eusebius. Others see it as a visual representation of salvation – what the land would look like if Ezekiel's vision, for instance, came to be. Art historian Beatrice Leal argues that the map was part of the decoration of a judicial hall rather than a church.[51]

The map has Jerusalem at its heart, but next to it is a relatively detailed depiction of the area around the 'Asphalt Lake', or the Dead Sea. Jericho is depicted with palm trees near it, as is Zoar, reflecting the crop's economic significance. Near the lower Jordan River there are also unidentified bushes which, some have argued, could be the mysterious balsam. Inside the Jordan there are fish swimming downstream, except the one closest to the Dead Sea, which has sensibly turned around and is swimming upstream, away from certain death in the lake. Nearby, there is a bridge crossing the Jordan and, on the opposite bank, a gazelle being chased, possibly by a lion. The map also depicts the hot springs of Callirrhoe. The Dead Sea itself is shown as one large lake. The Lisan Peninsula is conspicuously absent: with its highest point at -318 metres, it would have been visible even if the lake's level had risen to the maximum point suggested by archaeologists and proxy records: -387 metres. On the lake are two boats bearing different-coloured goods. The white mound on the ship closer to Jericho is likely a cargo of bulk salt. The second ship carries an ochre-coloured cargo which could be bitumen or grain. The four sailors on the two boats were defaced by later iconoclasts, who used mosaic stones to fill in much of the bodies of the destroyed figures.[52]

A SITE OF WEALTH, WAR AND REFUGE

There is archaeological evidence that goods and people were transported across the Dead Sea throughout the Hellenistic and Roman periods. At the northern edge of the lake there was a breakwater, which later became known as Rujum el-Bahr, with several mooring stones and the remains of a Hellenistic structure on it. When the water level dropped, the exposed stones of this artificial offshore anchorage provided shelter from wind and waves and allowed simple mooring by tying up directly to the mound. At high water levels, the submerged stone pile provided a firm grip for anchors.[53] In recent decades, the receding shoreline of the Dead Sea has revealed a number of anchors near Ein Gedi. One of these was a Roman-era wooden anchor with a lead stock. The anchor was fastened with ropes made of date-palm fibres, suggesting it was manufactured locally. Callirrhoe, which served as the port of Machaerus, a few kilometres uphill to the south-east, had a harbour platform.[54] The Madaba map suggests that the Dead Sea remained a bustling hub of maritime activity well into the sixth century.

A MOSAIC OF CULTURES

The Byzantine Empire, like the Roman Empire, Hellenistic kingdoms and Persian Empire before it, was composed of people of different cultures and faiths. Many cities across Byzantine Palestine had a hybrid population of Christians, Jews, Samaritans and pagans. For the Dead Sea area, too, we have evidence of a heterogeneous population.

Alongside the many Christian churches and monasteries north of the Dead Sea, there was also a Jewish synagogue at Naaran (the modern-day village of Nu'eima), north-west of Jericho. The mosaic that was uncovered in this synagogue bears a profusion of symbols, including two *menorahs* (candelabra) – the most popular Jewish symbol of late antiquity – and the zodiac, with a personification of each season as well as the chariot of the sun in its midst.

The latter is not a traditional Jewish symbol, indicating a degree of religious openness. In the late sixth or early seventh century, a new synagogue was established not far from Naaran, near Jericho's Tell es-Sultan. Often referred to as the Shalom al Israel ('Peace upon Israel') synagogue, because of a Hebrew inscription on one of its mosaics, it took a much stricter approach regarding ornamentation. Adhering closely to the biblical prohibition of graven images, it displayed only non-figurative mosaic decorations. Instead of including images in their zodiac mosaic to depict the different seasons and their labours, the builders of the new Naaran synagogue drew up a verbal list of them.[55] It seems that the Jewish tradition of having two nearby synagogues, one to pray in and another where never to set foot, has a long history.

Eusebius's fourth-century *Onomasticon*, a directory of place names in the Holy Land, tells us that Ein Gedi was a very large village of Jews. The settlement was re-established at some point after the Bar Kokhba Revolt. In the third century a synagogue was built at the centre of the village. The architectural remains of the village, the traces of extensive agriculture in the oasis and two hoards of gold coins from the fifth and sixth centuries all show that Byzantine Ein Gedi was prosperous.[56]

The area south and south-east of the Dead Sea had long been inhabited by Nabataean communities, as attested, for example, by the Nabataean temple of Khirbet et-Tannur, overlooking Wadi al-Hasa. In 1996 and 1997, excavations at Khirbet Qazone, just east of the Lisan Peninsula, revealed over 3,500 grave shafts. Most of these had been looted by tomb-robbers, but archaeologists were nonetheless able to identify a few *stelae* similar to the ones found in the former Nabataean capital at Petra. Further south, in Zoar, some 300 funerary *stelae* were uncovered, dating from between the fourth and sixth century. The majority of those buried were Christian. The names of people in the community derived from Nabataean, Arabic and Aramaic and attest to local conversions to Christianity. However,

A SITE OF WEALTH, WAR AND REFUGE

some 10 per cent of the graves belonged to Jews, with inscriptions in Aramaic rather than Greek, the language of the Christian epitaphs. Jewish *stelae* were often decorated with a *menorah*.[57]

Heterogeneous as it may have been, there were also tensions between different groups in the Byzantine period. In the sixth century, there was hostility between the Jews of Naaran and the non-Jews of nearby Jericho. These were possibly exacerbated by a Christian monk who decided to convert to Judaism, leave his monastery in Sinai and settle in the Naaran area, where he preached against Christianity and on behalf of his new creed. The community at Ein Gedi also exhibited suspicion towards outsiders, with the inscription on the floor of their synagogue warning 'anyone who slanders his mates in front of gentiles, or anyone who steals the property of a mate, or whoever reveals the secret of the town to gentiles'.[58]

In addition to tensions between them, communities around the Dead Sea were also vulnerable to raids by nomadic tribes. Through the development of a more efficient type of camel saddle, the nomadic tribes of the northern Arabian Desert gained new mobility and increased the distance of their raids. From the third century onwards, Greek and Latin sources speak of *saraceni* to describe this new danger.[59] Christian sources from the sixth century mention raids on monasteries in the Judean Desert and a monk who was killed while strolling along the Dead Sea. A conflagration destroyed Ein Gedi in the late sixth or early seventh century, though the question of whether this destruction was the result of a raid or was part of the persecution campaign against the Jews during the reign of either Emperor Justinian I (527–565) or Emperor Heraclius (610–641) remains unsettled.[60] The prosperity that the Dead Sea region enjoyed between 300 and 600 CE, and which was so vividly captured in the Madaba map, was beginning to draw to an end.

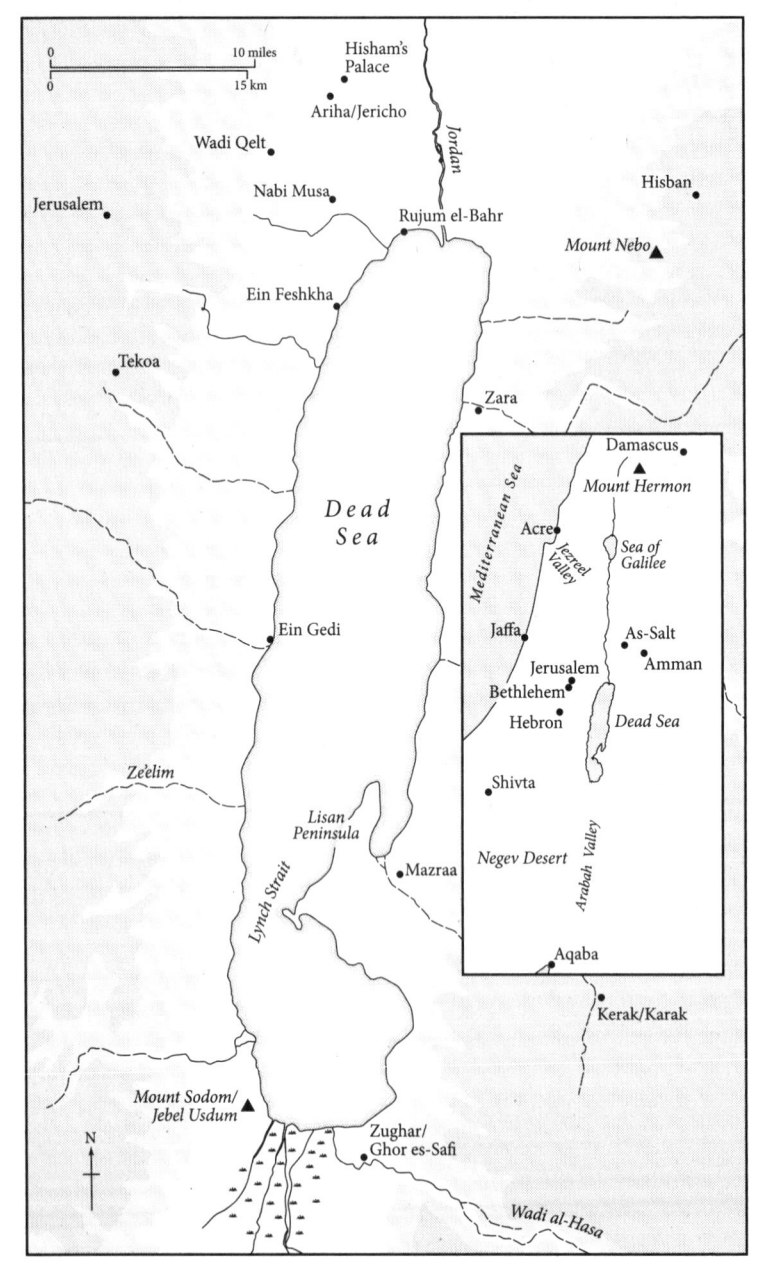

Map 3. Sites near the Dead Sea from the seventh to the nineteenth century CE.

CHAPTER FOUR

❧

A SITE OF DECLINE, NEW BEGINNINGS AND MYTHS

THE TENTH-CENTURY Jerusalem-born geographer al-Muqaddasi likened the area south of the Dead Sea to hell: 'verily this is a country that is deadly to strangers, for its water is execrable and he who should find that the Angel of Death delays for him, let him come here, for in all of Islam I know not of any place to equal it in evil climate.'[1] Jacques de Vitry, a theologian and chronicler who served as the Bishop of Acre in the early thirteenth century, also did not have many kind words for the Dead Sea. In his *History of Jerusalem*, he recounts how the Lord rained fire and brimstone upon Sodom and Gomorrah, whose people were exceedingly wicked and had 'sinned against nature'. Once the cities were destroyed, the Lord cast them into a bottomless pit where the lake now stands. While this lake had different names, de Vitry notes that it was often called the Devil's Sea.[2] The Dead Sea was already associated with the destroyed cities of the plain in antiquity, but it was during the medieval and early modern periods that authors increasingly described the area as cursed.

Clearly, the image associated with the lake had changed since the era of prosperity depicted in the Madaba map. From the sixth century onwards, the southern Levant underwent a number of political, cultural, social and climatic transformations. As a result, both the demographic composition and the economy of the Dead Sea region changed. For instance, balsam production ceased, while indigo and sugarcane were introduced to the area. However, because there is a dearth of sources for long stints between the seventh and eighteenth centuries, only an incomplete picture of the lake's history during this period can be reconstructed.

TRANSITION

After the prosperity of the Byzantine period came a time of decline. Archaeological evidence suggests that some settlements near the Dead Sea were abandoned abruptly in the late sixth and early seventh century. In others, activity continued, although with a marked decline, reaching a nadir around the tenth century. What brought about this decline? There are a number of suspects to consider.

The first of these is that the process of decline began already in the dusk of Byzantine rule in the southern Levant. Emperor Justinian I is known for his attempts to restore the empire. His long reign included Byzantine conquests in North Africa and in Italy and large-scale architectural endeavours such as the reconstruction of the Church of the Nativity in Bethlehem. However, the Byzantine Empire also experienced a number of shocks during his reign. In 536–537 there was a 'dust veil' event, in which the rays of the sun were partly blocked and the moon seemed dim. This led to stunted growth in plants that is recorded in tree rings.[3] In 541–542, bubonic plague hit the Mediterranean world. Likely carried by rats on ships crossing the sea, it killed up to one third of the population in certain regions. Famine often followed on the heels of plagues because, in times of outbreak, farmers were unable to cultivate their fields.

A SITE OF DECLINE, NEW BEGINNINGS AND MYTHS

However, in Palestine, urban centres as well as settlement on the fringe of the desert continued to grow in the second half of the sixth century.[4] While the bubonic plague undeniably created misery elsewhere in the Byzantine Empire, it is unlikely to have brought about the decline of the Dead Sea region and neighbouring areas.

One interpretation finds the chief reason for this decline in a change in the region's climate. As we have seen, between roughly 300 and 600 CE, the southern Levant probably benefited from wetter conditions, as attested by a rise in the level of the Dead Sea. The lake has two basins: a deeper and larger northern basin and a shallower and smaller southern basin. In periods of high precipitation, the Dead Sea covered both basins. In drier periods, the level of the lake fell. The two basins are separated west of the Lisan Peninsula by what became known in the nineteenth century as the Lynch Strait, which has an elevation of approximately -402 metres. Once the water level dropped below that height, the Dead Sea divided into two lakes. Such a drop likely coincided with the end of the Byzantine period in the southern Levant.[5] There is evidence to suggest that the seventh century saw a period of decreased sunspots, a solar minimum, which brought about cooler temperatures in parts of what we now call the Middle East. In the seventh century, agricultural activity declined not only around the Dead Sea but also in other areas of the southern Levant and as far north as Asia Minor. Pollen data indicate that, in certain areas, the cultivation of fruit-growing trees gave way to pastoralism. Severe and persistent droughts in the sixth-century Arabian Peninsula contributed to the downfall of the Himyar kingdom. However, the impact of the changing climate was not uniform across the whole Middle East and must not be overstated. For instance, the region known in antiquity as Decapolis, now in northern Jordan, flourished continuously from Roman times until the tenth century thanks to rainfed agriculture that produced considerable crop surplus.[6] Even near the Dead Sea, farming continued well after the end of Byzantine rule. Perhaps other factors were at play, alongside the drier climate?

Zooarchaeologist Guy Bar-Oz and his team study the demise of the thriving agricultural communities in the Negev Desert south-west of the Dead Sea at the end of the Byzantine period. Their excavations have not turned up evidence of mass disease. Instead, the ebbing and eventual abandonment of these villages in the sixth century happened chiefly, Bar-Oz believes, due to shifting trade routes. During the heyday of Byzantine rule, villages such as Shivta prospered due to the vineyards they cultivated. The wine they produced was traded across the Mediterranean. Furthermore, as landfill remains from this period illustrate, many exotic goods such as Red Sea fish passed through the settlements in the Negev. However, when demand for their wine fell, possibly because the population elsewhere in the Byzantine Empire declined, the local economy collapsed, and settlement activity soon ceased. It is possible that settlements near the Dead Sea were also adversely affected by shifting trade routes. Luxury Red Sea fish were brought to various sites near the lake – a distance of more than 180 kilometres – during the Byzantine period. However, these nearly disappear from the archaeological record from the seventh century onwards. In the region of Wadi al-Hasa, southeast of the Dead Sea, the Byzantine period saw a peak in population and desert agriculture, but here too it decreased thereafter. Brett Hill, who surveyed sites in the area, attributed the decline to a shift in the local economy away from long-distance trade through caravan traffic toward local exchange.[7]

Another factor to consider is the deterioration of security. In 614 the Byzantine southern Levant was overrun by the Persian-Sasanian Empire. While this temporary conquest brought about considerable destruction in Jerusalem and its environs, the Dead Sea area appears to have suffered little harm. A few monasteries in the Judean Desert were raided, though probably by Saracens

A SITE OF DECLINE, NEW BEGINNINGS AND MYTHS

rather than the Persian army.[8] Byzantine rule was briefly restored in 628/629 by Emperor Heraclius, but within a few years the whole region was conquered by a new force that, this time, came not from Mesopotamia and Persia in the east nor the Mediterranean world in the west, but from the Arabian Peninsula.

After the death of the Prophet Muhammad in 632, his successors spread from Arabia, defeated both the Byzantine and Persian-Sasanian empires and conquered much of what we now call the Middle East. Was the Islamic conquest responsible for population decline and the reduction in agricultural activity in Palestine? That this conquest roughly coincided with changes in land use in the southern Levant led some scholars to identify the former as having caused the latter. For instance, in 1967, geologists David Neev and Kenneth Orris Emery argued that, after the Arab conquest, overgrazing and the neglect of terraces led to the loss of much of the vegetation of the hills of Palestine. Moreover, they believed 'the highly-developed Roman and Byzantine irrigation systems in the entire Jordan Valley, still attested to by the abundant ruins of aqueducts near most springs, were neglected and destroyed by nature and man'. As a result, from the seventh century onward, runoff grew, as did the inflow of water into the Dead Sea, resulting in a rise of the level of the lake as well as an increase in its size.[9] This interpretation fits into a general tendency among many European and North American scholars in the twentieth century to see nomads as contributing to the degradation of a once-flourishing Roman-era environment across North Africa and the Middle East.[10] It is also reminiscent of the famous argument, put forward by the Belgian historian Henri Pirenne, that the Islamic conquest of North Africa led to the shrinking of trade across the Mediterranean and a broader economic decline in what used to be the Western Roman Empire.[11] A closer look at the Dead Sea area in the Early Islamic period reveals a more complex picture, one that includes both change and continuity.

THE DEAD SEA

CONTINUITY AND CHANGE UNDER THE UMAYYADS AND ABBASIDS

The initial period that followed the death of Muhammad is known as the Rashidun Caliphate, wherein the expanding Islamic empire was ruled by a number of *caliphs* – successors – who were elected or appointed from among tribal elites that were close to Muhammad. Then, one of these caliphs, Mu'awiya I (661–680), a former Governor of Damascus, broke with tradition and appointed his son as his successor. This led to the establishment of the Umayyad Caliphate, an Islamic dynasty that made Damascus their capital. The Umayyads invested heavily in grand construction projects around greater Syria, most famously the Dome of the Rock in Jerusalem and the Umayyad Mosque of Damascus. In 750 the Umayyad Caliphate was overthrown by the Abbasids, descendants of Muhammad's uncle. The Abbasids moved their capital to Baghdad. These geopolitical shifts in power left their mark on the Dead Sea region, where certain aspects of life continued, while others changed.

One notable change was that the region lost what had once been an important source of revenue and fame. By the Umayyad period the Dead Sea area had ceased to be a centre for cultivating balsam. Instead of Ein Gedi and Jericho, the widely recognised source of obtaining the 'true' balsam moved to a small garden in the Egyptian village of Matariyya, nowadays in north-eastern Cairo. At the end of the seventh century, the Bishop Zacharias of Sakha provided an account of how the garden had started, which gave it a holy pedigree: while in Egypt, Zacharias says, the infant Christ took sections of Joseph's staff and planted them in the ground. The fragments of wood grew to become the garden's balsam trees. Zacharias added that Joseph's staff was made from a tree that grew near Jericho. Myths aside, it is possible that part of the balsam plantation from the Dead Sea area, along with the secret knowledge of how to produce the perfume, was transported to Egypt in late antiquity. For almost a millennium, Matariyya was an integral part of a successful perfume industry. Alas, the last tree in the garden died following an inundation of the Nile in 1615, a loss that contributes to the mystery that still shrouds the 'true' balsam.[12]

Monastic presence in the Dead Sea area continued into the Early Islamic period. However, some of these monasteries shrank in size and population. Some, like the small monastery built during the Byzantine period atop Masada, were abandoned already in the seventh century. Larger monasteries or those with good access to water and agricultural goods continued to function into the eighth and ninth centuries, especially in the area between Jericho and the Dead Sea. However, the number of active sites in places like Wadi Qelt diminished in the Umayyad period, shrinking further under the Abbasids. The declining volume of Christian pilgrims from abroad meant that the monasteries lost a major source for donations.[13]

Not every aspect of life was marked by decline. The clearest evidence for innovation in the period that followed the Islamic conquest can be found in the palatial estate that the Umayyads established some 4 kilometres north of modern-day Jericho. The site is known today either as Khirbet al-Mafjar or, colloquially, as Hisham's Palace. It earned the latter name because of two of the inscriptions found at the site which mention the Umayyad Caliph Hisham ibn Abd al-Malik (724–743). The complex included a palace, mosque, audience hall, bath and pavilion, all enclosed within a wall with two gates: one leading north and the other south to Jericho.[14] The palace building itself boasted an elaborate water drainage system and beautiful mosaics, including the famed tree of life. This mosaic depicts a large fruit-bearing tree with two relaxing gazelles on one of its sides and a lion ferociously attacking a third gazelle on the other. As we have seen, Jericho had a long tradition of lavish architecture. Hisham's Palace was the latest in a long line.

In 749, a major earthquake struck the entire Jordan Valley. The earthquake may have even triggered a tsunami in the Dead Sea.[15] Hisham's Palace and other buildings in Jericho and on the eastern side of the lake were severely damaged. However, the earthquake did not bring the Early Islamic settlement in Jericho to an end. The site at Khirbet al-Mafjar was reoccupied during the Abbasid period. Just north of the Umayyad palace, archaeologists found the remains of an Abbasid agricultural estate with a *khan* – a caravanserai or roadside inn – next to it.[16]

The Muslim newcomers were, at first, a minority. We do not have accurate population figures, but Christians probably constituted the majority in Palestine until at least the eleventh century.[17] In Jericho, Christians and Jews continued to coexist alongside the new Islamic elite. A basilica that stood some 3 kilometres east of Jericho, at Khirbet en-Nitla, was rebuilt a number of times. Originally constructed during the Byzantine period, it continued to function until the ninth century. About halfway between Jericho and the Jordan River, a small chapel with a Syriac inscription, probably a Nestorian hermitage, was only constructed in the mid-seventh century. A rare glazed, lidded bowl originating from Iraq and found at the site demonstrates that it was active well into the eighth.[18] Early Islamic coins were found in excavations in the Shalom al Israel synagogue near Tell es-Sultan in Jericho. It has been suggested that this synagogue served Jewish tribes that were expelled from Medina in the Arabian Peninsula around the time of the Prophet Muhammad and settled in Jericho, among other places. The synagogue continued to function until the eighth century, when it was abandoned, possibly because of the earthquake of 749.[19]

East of the Dead Sea, there was visible damage to the monastic complex on Mount Nebo, likely caused by the same earthquake.

A SITE OF DECLINE, NEW BEGINNINGS AND MYTHS

Some repairs were carried out afterwards, with archaeologists finding a few sherds that date to the Abbasid period. However, the site appears to have fallen out of use shortly thereafter.[20] The Sanctuary of St Lot near Zoar also remained active long after the Islamic conquest, as attested by a mosaic with an inscription dating to 691. The decline and final abandonment of this site probably took place in the early ninth century, during the Abbasid period, as suggested by a jug with a unique Kufic (early Arabic script) inscription on its shoulder that was found in the uppermost level during excavations. Based on the lack of destruction at the site, archaeologist Konstantinos Politis believes it was abandoned peacefully rather than as a result of an earthquake or a raid.[21] While we cannot be certain that it was caused by it, the abandonment of settlements near the Dead Sea coincided with power shifting, under the Abbasids, from Damascus to Baghdad far in the east.

FATIMID, CRUSADER AND MAMLUK SOCIOECONOMIC TRANSFORMATIONS

After the decline in Abbasid power, control over the southern Levant passed between a number of rulers of different faiths. In the late tenth century, the North African Fatimid Caliphate expanded and seized control of Egypt, where they soon established their new capital, Cairo. From their base in Egypt, the Fatimids maintained a shaky hold over Palestine as well, though their rule there was challenged by the Byzantines, by local warlords and eventually by the Seljuk Empire that, for a short while, stretched from Central Asia to the Levant. Then, at the end of the eleventh century, a new force arrived from Europe: the Franks of the First Crusade. Among other states, the Crusaders established the Kingdom of Jerusalem, which lasted, in its original form, from 1099 to 1187 and ruled over both sides of the Dead Sea. The

THE DEAD SEA

Kingdom of Jerusalem was eventually defeated by Salah al-Din, or Saladin, who founded the Ayyub Dynasty. Although the Third Crusade managed to reestablish Frankish rule on the coast of Palestine, the Dead Sea remained in Ayyub hands until the dynasty was overthrown by an uprising of its own former slave-soldiers, the Mamluks.

While these political changes certainly left their mark on the Dead Sea region, the economy of the area followed a different rhythm. Salt mining around the lake continued as it had done in the Early Islamic period. From the tenth-century Jerusalem-born physician Muhammad ibn Sa'id al-Tamimi we know that various types of salt from the vicinity of the Dead Sea were marketed across al-Sham (Greater Syria or the Levant). These were distinct from each other by virtue of their brittleness, colour and purity. For instance, he tells of rock salt with a blue hue which likely came from Mount Sodom, south-west of the lake. He disputes the Roman-era physician Galen, who had said that only bitter salt could be found around the Dead Sea. Instead, al-Tamimi mentions white, edible and even pleasant-tasting salt that came from the western shores of the lake. Some of this salt was collected by the people of Jericho, sold in Jerusalem and used for cooking and seasoning. Salt was also mined near Zara, a settlement on the eastern shore of the Dead Sea where Roman-era Callirrhoe once stood.[22]

Two new crops were introduced to the southern Levant in the Early Islamic period: sugarcane and indigo. Of the two, indigo (*Indigofera tinctoria*, also known as True Indigo), from which deep blue-coloured dye was produced, seems to have been grown on a commercial basis near the Dead Sea earlier on. In antiquity, indigo had been exported from India to the Mediterranean world, where it was used, in powdered form, in paint and cosmetics. The tenth-century Islamic geographer Ibn Hawqal mentions Zughar – the Arabic name for medieval Zoar – as a centre for production and

A SITE OF DECLINE, NEW BEGINNINGS AND MYTHS

considerable trade in indigo. The Jerusalem-born geographer al-Muqaddasi, a contemporary of Ibn Hawqal, described the Jericho region as a land of excellent indigo and date palms, where bananas and fragrant flowers were also grown. Indigo was still the principal crop of Jericho and the Jordan Valley in the mid-twelfth century, according to the geographer Muhammad al-Idrisi. Because the mill technology used in the production of indigo and sugar is similar, it is possible that some of the earlier sites near Zughar associated with sugar production were initially used to process indigo leaves.[23]

Sugarcane is indigenous to India and South East Asia. It was introduced to what we now know as Iraq before the Arab conquest in 637. From there it spread to Syria, the southern Levant and Egypt after the Islamic conquest, at some point before the tenth century. The cultivation of sugarcane requires abundant sunlight and water for irrigation. Furthermore, in the medieval Levant, the production of sugar depended on fast-flowing water to power the mills. Very warm and stream-fed sites in the Jordan Valley and near the Dead Sea were deemed by contemporaries as suitable for this crop, and these became important hubs for this new industry.[24]

There is some debate about when precisely the sugar mills of Jericho and Zughar were first established. Sites that by their very name preserved the memory of sugar production – Tawahin as-Sukkar ('sugar mill') – can be found next to both towns. There may have been some small-scale production during the Fatimid period. On arrival in the Levant, many of the Crusaders encountered sugarcane for the first time, dubbing it 'honey-cane' or 'wild honey'. They chewed it to suppress their hunger when they first marched into the region. Within a few decades the Franks enthusiastically expanded the production of sugar, which became a key cash crop in the economy of the Latin east. Sugarcane production in the vicinity of the Dead Sea probably reached its peak under the Mamluks. In the early thirteenth century the geographer

Yaqut al-Rumi al-Hamawi noted that sugarcane was grown throughout the Jordan Valley, including in Jericho. His contemporary the Bishop of Acre Jacques de Vitry described the valley in very evocative terms: 'The fields by the riverside drop sweetness from the thick multitude of sugar-canes, and yield abundance of sugar.'[25] The traveller Jacob of Verona, who visited the country in 1335, mentioned sugar plantations east and south of the Dead Sea. The Dominican friar Felix Fabri, to whom we will return shortly, alluded to sugarcane still growing by the Jordan River in the 1480s. However, the sugarcane industry in the Levant declined by the late fifteenth century as high-quality production of sugar was beginning to take off in places like Portuguese Madeira. Soon, many of the forty-three sugar mills that archaeologists know of in the Jordan Valley and near the Dead Sea were abandoned.[26]

As we know well from sugar plantations in the Caribbean in later centuries, growing and producing sugar was very labour-intensive. There is some evidence to suggest that enslaved people from Africa were brought to the Dead Sea area for this purpose, but it is not clear when precisely this happened.[27] Excavating in Ghor es-Safi (historic Zoar/Zughar), Politis observed that, sometime in the late eighth or early ninth century, new people who used a different type of architecture settled in the area. Crude and modest walls, representing dwellings for a poorer population, replaced the arched doorways, columns and other luxurious attributes of earlier periods. The Fatimids made use of enslaved people from Nubia and the Fezzan in their military, and also for agricultural work in Egypt.[28] This latter practice may have been extended to the Dead Sea area. In his account of Zughar in the second half of the tenth century, al-Muqaddasi describes the local residents as 'black-skinned and thick-set'.[29]

Later medieval authors also encountered darker-skinned residents at both Zughar and Jericho. One of these was the chronicler Fulcher of Chartres, who accompanied Baldwin I, the Frankish

A SITE OF DECLINE, NEW BEGINNINGS AND MYTHS

King of Jerusalem (1100–1118), when he first explored the southeastern parts of his domain. Like Muqaddasi before him, Fulcher gives us an eye-witness account of Zughar, which he calls Segor. He describes 'a village most favorably situated and abounding in the fruits of the palm which they call "dates" and which we ate all day for their pleasant taste'. When it comes to the local population, Fulcher was far less generous: the 'Saracen' (Muslim) inhabitants had fled when they learned the Franks were approaching, leaving behind 'some who were blacker than soot'. To this he added: 'these we left there, despising them as if they were no more than sea-weed.'[30] The twelfth-century English historian William of Malmesbury embellished further on Fulcher's account. He described those left behind as 'Ethiopians', whom 'our men thought it unworthy of their mettle to put to death, regarding them as objects of humor rather than hostility'.[31]

The ethnic diversity of the Dead Sea region had clearly increased. In documents conferring land to the Frankish elite, we have an early mention of another group that resided in the Dead Sea area: the Bedouins. In 1138 King Fulk of Jerusalem and his wife, Queen Melisende, granted the canons of the Church of the Holy Sepulchre in Jerusalem the *casale* – the village with its attached lands – of Thecua (Tekoa). The grant included not only the land but also the Bedouins that inhabited it. Nomadic groups of Saracens had been a feature of the Dead Sea region for centuries. By the eleventh century, Arabian tribes had moved into certain areas of Palestine and Syria. The Franks used the term 'Bedouins' loosely, describing in this way the Muslim and Christian Arabic-speaking pastoral nomads who inhabited the deserts to the east and south of the Latin Kingdom.[32] In later centuries, as we will see, Bedouin tribes would become the most dominant group in the lake's vicinity.

The Fatimid and Crusader periods have not provided us with a trove of documents such as the ones left behind by Babatha, shedding light on the everyday life of people near the Dead Sea. However, Frankish documents and accounts by travellers do tell us something about how the economy in the region worked. For instance, the Frankish rulers in Jerusalem allowed the inhabitants of Thecua to continue gathering bitumen from the Dead Sea and salt from the lake's shores. The eleventh-century Persian poet and traveller Nasir Khusraw reported that bitumen from the Dead Sea was traded across several countries and used to protect the trunks and roots of trees from worms and other 'things that creep below the soil'.[33]

A Frankish charter from 1152 also reveals that boats continued to transport goods across the Dead Sea. Moreover, the transport of such goods was taxed, with some exemptions.[34] We know of maritime activity, crossing the Dead Sea between Zughar in the south and the coast near Jericho in the north, also from the twelfth-century Muslim geographer Muhammad al-Idrisi. Furthermore, in 2006, an archaeological team led by Asaf Oron found four unique composite anchors along the north-western shores of the Dead Sea. Two of these were revealed by the retreating shoreline near Khirbet Mazin. Preserved in their entirety, each of the two anchors included a wooden shank and arms, a stone weight and a rope. Judging from the grey lime plaster or mortar remains found on one of them, the stone weights were likely taken from the ruined fortifications of the Hasmonean period at Khirbet Mazin. Carbon dating based on the rope and wood suggest that the anchors were used between the tenth and twelfth centuries, that is, the Fatimid and Crusader periods.[35]

Further Frankish sources reveal that the Latin patriarch of Jerusalem owned immense tracts of land in Jericho. One interesting facet in such documents is what they tell us about how fresh water drawn from the springs in the vicinity was divided between different users. Eustache Grenier, who received land along with a mill in Jericho as a wedding present from the

Patriarch of Jerusalem, was entitled to receive water from a local spring once a fortnight for twenty-four hours. After Grenier's death, the amount of water allotted to the mill was doubled, so that it flowed for a whole day once a week. This method of water distribution, based on time shares rather than quantity, remained a characteristic of Jericho for a very long time. In the 1990s more than 900 people had time-based claims to the water coming out of the spring of Ain es-Sultan. The municipality of Jericho employed twelve officials, whose job it was to coordinate the distribution of water through various canals. Their responsibilities included blocking and unblocking sections of the canals according to a calendar that represented water rights.[36]

NATURE, RELIGION AND POWER

Both Muslim and Christian authors commented on the unique topography and environment of the Dead Sea. The tenth-century Islamic geographer Istakhri, for instance, understood perfectly well the hydrology of the lake's catchment, centuries before the arrival of modern European and American explorers. He noted that the Jordan Valley was lower than the rest of Syria and Palestine, and that the Jordan River flowed to the lowest point, the Dead Sea, which he called 'the Stinking Lake' (*al-Buhairah al-Muntinah*).[37] Furthermore, medieval writers were aware that the level of the lake changed over time. Writing in the late tenth century, al-Tamimi says: 'reliable men, among them elders of Zoar and the elders of Jericho, claim that the waters of this sea over the course and length of years are deficient in a patently noticeable way.' These elders testified that, compared to the days of their fathers and grandfathers, 'the waters of the sea are dwindling'. This observation correlates with evidence from across the eastern Mediterranean indicating that the area suffered from drier than average conditions during the late tenth and first half of the eleventh century.[38]

THE DEAD SEA

In the Crusader period, Fulcher of Chartres was fascinated by the uniqueness of the Dead Sea area. He tasted the lake's water and found it 'more bitter than hellebore'. Fulcher observed that next to the Dead Sea was 'a great and high mountain of salt', presumably Mount Sodom. He wondered whether the saltiness of the lake was caused by the waves brushing against the mountainside and the rain washing salt down from it. He also hypothesised that the lake was somehow connected underground to 'the Great Sea', that is, the Mediterranean, thereby contributing further to its saltiness.[39]

As well as being the subject of intellectual curiosity, water also played an important spiritual role. After the Crusader conquest of Jerusalem, Frankish nobles and their men came down to bathe in the Jordan at the traditional baptism site near Jericho. Notable authors from this period such as William of Tyre and Burchard of Mount Sion described in their writings the biblical significance of the landscape north of the Dead Sea and of the lake itself. Speaking of the Jordan River, Jacques de Vitry noted how 'our Redeemer, having been baptized by St. John in that river, hath sanctified this river by the touch of His most pure flesh, and hath bestowed regenerative power upon all its waters'.[40]

The religious significance of the area brought about a modest revival of monastic activity in the Crusader period. In the twelfth century, Latin presence was established in the former Byzantine monastic cells on Mount of Temptation, and Orthodox monasteries were established near Jericho. However, security remained a problem. The Orthodox monastery of St John beside the Jordan River came under attack in 1139, and six monks were decapitated. To prevent such attacks, all the monasteries had to be fortified. Protecting pilgrims on the road from Jerusalem to Jericho and those continuing to the Jordan River became one of the tasks of the Templars, a monastic military order established in 1119. The German pilgrim Theoderic, who visited the area in 1172, says the

Templars used the caves on Mount of Temptation as one of their bases.[41]

The area near the Dead Sea played a role in the greatest defeat of the Crusader Kingdom of Jerusalem. The Franks called their territory east of the Jordan River and the Dead Sea 'Oultrejordain'. From its construction in 1142, Kerak (modern-day Karak), with its formidable 'fortress of the raven', served as the administrative centre of Oultrejordain. As the crow flies, Kerak is only some 16 kilometres east from the shore of the southern basin of the Dead Sea. However, as it is perched on a hilltop, to reach Kerak one has to ascend more than 1,300 metres above the level of the lake. Kerak overlooked the King's Highway, which connected Damascus with either Egypt in the south-west or the holy cities of Mecca and Medina in the south-east. It was this latter vantage point that was of great interest to the last Crusader Lord of Oultrejordain, Reynald of Châtillon, a man who is remembered in Islamic chronicles as 'one of the most devilish of the Franks'.[42] Having already gained a reputation for cruelty as Prince of Antioch, in 1176 he married Stephanie de Milly and through her gained possession of Kerak. Reynald used this stronghold to stage raids on Muslim pilgrim caravans en route to Mecca. He also sent a naval expedition down the Red Sea. This angered the Sultan of Egypt and Syria, Saladin, who twice laid siege to Kerak, in 1183 and 1184, but failed to take it. Undeterred, Reynald continued to harass Muslim caravans despite a temporary truce between Saladin and Baldwin IV, King of Jerusalem. This last attack, the chroniclers say, led Saladin to launch his campaign that culminated with the defeat of the Frankish army at the Battle of Hattin in 1187.[43]

The next major impact on the Dead Sea area was that of the Mamluks, who took over the southern Levant in the later thirteenth century. Their impact owed much to a transformational figure with an incredible life story. Baybars was a Kipchak Turk who grew up on the steppe north of the Black Sea. After his country was overrun by the Mongols in the 1240s, his tribe sought refuge in northern Bulgaria from where some of them were later transported to Anatolia and sold as slaves. At first it proved difficult to find a purchaser for Baybars, who was tall and young but had a white speck in one of his blue eyes, which some saw as a bad omen. Eventually, he ended up serving as a Mamluk (slave-soldier) in the military of the Ayyub Sultanate in Egypt. Baybars displayed outstanding military skills, earning a reputation as an ambitious, effective and ruthless warrior. He was eventually emancipated and appointed as one of the commanders of the Sultan's army. Growing in strength, a group of commanders murdered the Sultan and established a new Mamluk Sultanate. Although he participated in this coup, subsequent power struggles in Egypt forced Baybars to flee to Syria where he bided his time. His opportunity to rejoin the Mamluk forces came thanks to the appearance of a formidable external foe, the Mongols, who had captured Baghdad and were poised to conquer the Levant. In 1260 Baybars distinguished himself in the Battle of Ain Jalut, where the Mamluks defeated the Mongols, halting the latter's advance in the Middle East. Shortly after the battle, Baybars assassinated the Mamluk Sultan Qutuz and seized the throne. He remained sultan until his death in 1277. During his reign, he waged war on the strongholds of the Franks on the Mediterranean coast and further inland, greatly reducing the area under their control. Baybars was also a great builder, who established and repaired mosques, bridges and forts throughout his realm. In the vicinity of the Dead Sea, he refortified Kerak in the east, As-Salt in the north and, possibly, Hisban in the north-east, all with the aim of securing inland routes against both Frankish and Mongol raids.[44]

As we have seen, Baybars ordered the construction of a *maqam* at Nabi Musa, the site associated with the tomb of Moses, south of Jericho, in 1269. When constructing this shrine, his workers

A SITE OF DECLINE, NEW BEGINNINGS AND MYTHS

made use of Frankish columns, capitals and friezes. In doing so, the Mamluks sought to convey a declaration of military victory and supremacy. Alongside the architectural message of dominance, an inscription at Nabi Musa includes Baybars's titulature: 'the Sultan of Islam and the Muslims, master of kings and sultans, conqueror of the cities, destroyer of the Franks and Mongols, who wrests castles from the hands of unbelievers, heir to kingship, sultan of the Arabs, Persians and Turks, the Alexander of [his] time ...'.[45] The shrine at Nabi Musa demonstrated the might of Baybars and the Mamluks, as well as the religious significance of the area near the Dead Sea for Muslims.

Mamluk rule in the southern Levant coincided with favourable climatic conditions. Paleoclimatic research indicates that the Mamluk period saw slightly higher precipitation. After falling around the eleventh century, the level of the Dead Sea rose, covering the Lynch Strait and connecting the lake's northern and southern basins. These wetter conditions enabled effective irrigation of sugar plantations and the operation of mills in Jericho and Zughar. The favourable climate also helps to explain why archaeologists have been able to find Mamluk-period artefacts in places like the Ein Gedi oasis and the Murabba'at caves after a long gap since late antiquity.[46]

However, this period of enhanced agricultural activity was short-lived. Pollen analysis indicates that the higher precipitation of the thirteenth and fourteenth centuries decreased in the fifteenth century, likely affecting water-intensive crops such as sugarcane adversely. The demise of sugar production was symptomatic of a broader social and economic crisis in the southern Levant in the late Mamluk period. Mismanagement and financial turmoil in Cairo, coupled with diminished trade in the eastern Mediterranean, increased Bedouin raids and the recurrence of plague, led to a decrease in the region's population. Some administrative centres were abandoned by the sultanate's central authority, while a number of villages were vacated by their residents.[47]

In 1516 the outdated Mamluk army was defeated by the firearm-wielding Ottomans. The southern Levant subsequantly experienced a brief economic upswing, benefiting from grand projects such as the construction of the walls of Jerusalem by Sultan Suleiman the Magnificent (1520–1566). Data from Ottoman records suggest that the population of the region grew in the late sixteenth century and the number of unoccupied villages fell. However, over time, the southern Levant and especially the Dead Sea area became a peripheral and insignificant part of the Ottoman Empire. The lake area was dominated by Bedouins, and the population of settlements shrank.[48] While it still figured in the writings of pilgrims, in the European imagination the Dead Sea became shrouded in mystery and myth.

A SITE OF MYTHS

An unsigned twelfth-century text in Latin, *Descriptio locorum circa Hierusalem adjacentium* ('A description of the places adjacent to Jerusalem'), makes an interesting though unfounded assertion: not only were there no living creatures inside the Dead Sea but any bird flying over the lake would drop dead almost immediately because of its poisonous vapours. The belief that birds could not fly over the lake persisted for centuries.[49] In the medieval and early modern periods, the lake, sometimes called 'the Devil's Sea', developed a PR problem, especially in Europe.

Pilgrims and travellers visiting the Dead Sea area received much of their information (and misinformation) from local guides. Take, for example, Rabbi Benjamin of Tudela, who travelled from the Iberian Peninsula to the Persian Gulf and back in the second half of the twelfth century. While he only appears to have viewed the Dead Sea from a distance, standing atop Mount of Olives east of Jerusalem, he was nonetheless able to tell his readers about the salt pillar that was once Lot's wife near the lake. Benjamin relays that, although sheep continually licked the pillar, it miraculously grew again and retained its original shape. When Burchard of Mount Sion, a German Dominican friar who toured the area in the late thirteenth century, wanted to see the pillar of salt, locals warned him that the place was unsafe on account of wild beasts, serpents and worms, and especially because of the Bedouins who dwell in those parts. The latter, he was told, were exceptionally bold and evil men.[50]

One of the chief contributors to the European image of the Dead Sea as a mythical space was Felix Fabri. Born in Zurich, Fabri became a Dominican friar and travelled extensively in the Holy Land in the early 1480s. While in Jerusalem, Fabri and his fellow pilgrims wanted to travel down to the Jordan River and see the Dead Sea. The local Muslim guides tried to dissuade them from going there: while the Jordan was blessed, they said, the Dead Sea had been cursed by God. Like Burchard before them and so many later foreign travellers up to the early twentieth century, the pilgrims were also warned that the journey would be unsafe because of the dangers posed by Bedouin marauders. The guides told Fabri and his companions of the many harmful creatures that lived along the shores of the lake: poisonous animals, small and large, such as lions, bears, boars and serpents. One poisonous and 'noble' serpent-like monster, known as Tyr or Tyrus, attracted Fabri's attention. Not found anywhere else in the world, this serpent had poison that, when extracted, provided a key ingredient for tyriac (or theriac), a most powerful and precious drug. According to Fabri's guides, so valuable was this tyriac that the Mamluk Sultan had forbidden foreigners to go near the Dead Sea lest they steal the Tyr.[51]

Fabri describes the Tyr as being about 50cm long but only as thick as a thumb. Its colour was yellowish with some tinges of red. Though not very big, it was surprisingly powerful and characterised by fits of rage. When angry, the Tyr flashed a fiery tongue, whirled around very quickly and slid with a dreadful hiss while its body glowed like red-hot iron. At such times, its small head swelled until it was bigger than its body. The Tyr's poison caused the body of its victim to swell and burst. There was no remedy against its bite unless one amputated the bitten organ. Fabri goes as far as saying that its poison was so strong that should the Tyr bite a horse, its rider too would take the poison and die.[52]

Though some of Fabri's details are clearly far-fetched, the practice of using serpents to produce drugs and antidotes is well known since antiquity. The late Roman physician Galen wrote an entire treatise about the cure-all theriac, which was usually made from a mix that included opium and viper's flesh. From medieval writers we know that snakes in the Levant, and especially near the Dead Sea, were used for this purpose. For instance, the Arab geographer Muqaddasi, writing in 985, mentions the serpent 'tariyakkiyah' that was found near Jericho. Out of its flesh the excellent 'Tariyak of Jerusalem' was produced, an antidote against poisoning that was considered to be without equal. In the thirteenth century, Jacques de Vitry also described how the Tyr's flesh was used for the preparation of theriac. Found near Jericho by the fords of the Jordan, it served as a universal panacea against all kinds of poisons. Setting the scene for Fabri's fanciful portrayal of the Tyr two centuries later, de Vitry says this creature had hooked legs and a long tail, the features of a lizard and a skin like that of a crocodile.[53] From medieval times to the early modern period, the image of the Dead Sea area was suspended between the real and the imagined.

Sixteenth-, seventeenth- and eighteenth-century Greek Orthodox *Proskynetaria* – a term used for both travel guides for pilgrims and canvases depicting the sacred landscape of the Holy Land – portray the Dead Sea in either negative or peripheral ways. The *Proskynetaria* focus on those places in the sacred landscape

A SITE OF DECLINE, NEW BEGINNINGS AND MYTHS

that would have been central to the pilgrim's journey: Nazareth, Bethlehem, Jerusalem and baptism in the River Jordan. Those canvases that show the Dead Sea relegate it to the background.[54] The written travel guides, for their part, repeat many of the by now familiar tropes associated with the lake. They present the Dead Sea as a cursed place because of the biblical destruction of Sodom: 'through this sea God makes manifest both Hell and the malice of Sodom, so that we may see and hear and become prudent in ourselves.'[55] The guides go on to enumerate the sexual sins that one must avoid. Much like earlier texts, the travel guides explain that the cities of the plain had sunk under the surface of the Dead Sea. Pilgrims and travellers visiting the lake often complained that it stank of sulphur. One sixteenth-century Greek Orthodox author claimed that the lake smelled like a Jewish woman. Some of the information the *Proskynetaria* provided is less contentious: the lake does not have anything living inside it and occasionally it emits tar (bitumen), which could be used to protect vineyards from caterpillars. However, the authors, who may not have visited the area themselves, struggled to describe the accurate size of the lake. One early-seventeenth-century book reckoned the way around the Dead Sea measured some 700 miles. Another put it at 300 miles and a third, slightly more accurately, at a mere 100 miles.[56]

The difficulty to estimate the size of the Dead Sea is reflected in late medieval and early modern visual representations of the lake. These tended to emphasise the site's religious importance rather than provide an accurate representation of its size and scale. The *mappa mundi* preserved at Hereford Cathedral in the west of England, a map of the known world prepared around the year 1300, shows the Dead Sea with Sodom and Gomorrah half-sunk in the middle of it. To the left stands Lot's wife, looking at the cursed cities and turning into a pillar of salt. Because its primary objective was religious instruction, in terms of geographic proportions, the *mappa mundi* is misleading. Jerusalem, which stands at the centre of the map, is quite far from the lake even though, in reality, they are only some 25 kilometres apart. On the map the Dead Sea appears nearer to Babylon and the Euphrates, whereas in fact these are hundreds of kilometres away.

By the sixteenth century, cartography across Europe was beginning to transform, giving maps greater equality of scale, distance and orientation.[57] The maps produced by Thomas Fuller in 1650 provide a more realistic estimation of the relative distance between various sites across the Holy Land and pay much closer attention to the area around the Dead Sea. Their main purpose, however, remained religious instruction. Fuller therefore plotted on the maps in his book several sites and stories that were mentioned in the Bible, such as Moses standing atop Mount Nebo, the Children of Israel crossing the Jordan River, and the walls around Jericho. The maps reflect well the uncertainties that persisted until archaeological excavations got underway about the location of several biblical sites. For instance, Fuller places Zoar and Lot's cave near the north-western shore of the Dead Sea, above Ein Gedi, rather than on its south-eastern edge. The map of the Dead Sea offers a very detailed layout of biblical settlements west and north of the lake, but none appear on its eastern side, which is conveniently covered by a dedication written by Fuller and a coat of arms.[58] As Joan Taylor points out, with the abandonment of most of the settlements around the Dead Sea and due to security concerns, most of the pilgrims and travellers after the Frankish period only visited the northern edge of the lake, and did not venture further east or south. Hence, it is not surprising that, during the early modern period, sites near the more accessible north-western shore became identified with biblical settlements. Rujum el-Bahr, the artificial mound of stones just off the northern shore of the lake, became associated with Zeboyim, one of the destroyed cities of the plain. The ruins of Qumran were believed to be the remains of Zoar.[59]

A SITE OF DECLINE, NEW BEGINNINGS AND MYTHS

From the turn of the eighteenth century onwards, we begin to see travellers who questioned some of the received knowledge on the Dead Sea. For instance, the English scholar and clergyman Henry Maundrell dispelled the myth that birds could not survive flying over the lake. He attests to seeing 'several birds flying about and over the sea without any visible harm'. Maundrell tested the buoyancy of the lake, noting that the water 'bore up my body in swimming with uncommon force'. He did not find any bitumen at the place that he visited along the shore but reported having seen several lumps of it that had been brought up to Jerusalem.[60] Other European travellers who visited the lake in the following decades described how the Arabs gathered asphalt from the shore in 'considerable quantities every autumn', and sold it in Egypt where it was used for dying wool.[61] Clearly the trade in Dead Sea bitumen continued into the Ottoman period.

Another eighteenth-century English clergyman and traveller who adopted a more scientific approach when exploring the Dead Sea area was Richard Pococke. Benefiting from a classical education, Pococke was familiar with the estimates of Diodorus, Pliny and Josephus as to the size of the lake. However, he tried to make his own estimates according to what he could see. Pococke, who visited the Dead Sea area in the late 1730s, describes how the Bedouins used evaporation pools to extract salt: 'The Arabs make pits on the side of the lake, which are filled by its overflow on the melting of the snow, and when the lake is lower, the water evaporates, and leaves a cake of salt, which is about an inch thick.' In all likelihood, the same method to produce salt had been used around the lake for several centuries. Pococke was warned by local monks of the bad air near the lake and told of a Carmelite monk who had died after visiting it. Nonetheless, Pococke took a sample of the water to have it analysed, tasted it and bathed in the lake. Two days later he was seized with an 'extraordinary disorder of the stomach' alongside 'a very great giddiness of the head'.[62] Pococke's approach to investigating the area was considerably more empirical than Fabri's had been a few centuries earlier. The scientific exploration of the Dead Sea, along with a better understanding of why its waters must not be consumed, continued to develop in the nineteenth and twentieth centuries.

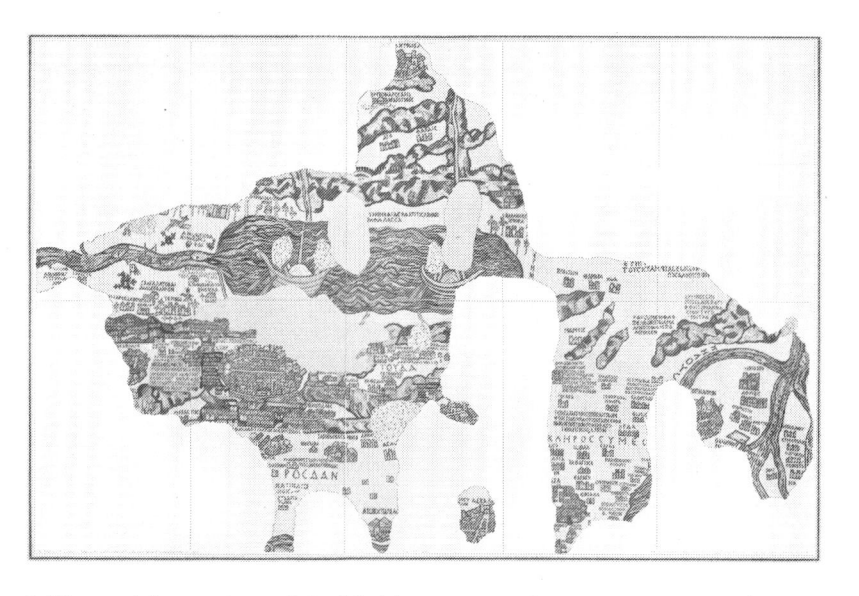

1. The surviving sections of the Madaba map, a sixth-century CE mosaic that was discovered at the Greek Orthodox Church of St George in Madaba, Jordan. The map offers a glimpse into life in the Dead Sea area during the Byzantine period.

2. A section of the map *A Pisgah Sight of Palestine*, prepared by Thomas Fuller in 1650. Focusing on the biblical importance of the country, Fuller shows the four destroyed cities of the plain, with flames still rising from them, in the middle of the Dead Sea.

3. Tell es-Sultan, the mound that was once ancient Jericho, during the excavations of Carl Watzinger and Ernst Sellin in the first decade of the twentieth century. The northern edge of the Dead Sea can be seen in the right-hand corner.

4. The excavations led by Carl Watzinger and Ernst Sellin at Tell es-Sultan, overlooking the Jordan Valley.

5. Qasr al-Abd ('Castle of the Slave'), an unfinished monumental structure built by the Tobiad family north-east of the Dead Sea in the Hellenistic period.

6. The Roman-era fortress of Machaerus, east of the Dead Sea. Originally built by the Hasmoneans, it was expanded by Herod and is believed to be the site where John the Baptist was executed. The fortress was destroyed during the First Jewish–Roman War.

7. The view from the eastern side of Masada in the early twentieth century: the remains of walled Roman camps, dating back to the siege around the fortress in 73/74 CE, the Lynch Strait and the Lisan Peninsula.

8. The palm tree 'Methuselah' at the Arava Institute for Environmental Studies in Ketura. It was germinated from a Roman-era date seed that was uncovered during archaeological excavations on Masada.

9. The Greek Orthodox Monastery of the Temptation on the cliffs of Mount Quruntul, overlooking Jericho. This mountain was once the site of the Hasmonean fort of Dok. Its caves were used by hermits in the Byzantine and Crusader periods.

10. The tree of life mosaic at Hisham's Palace, north of Jericho. The palace was built in the first half of the eighth century CE during the Umayyad period.

11. William Holman Hunt's *The Scapegoat*. Painted in the 1850s, it shows a goat, which had been sent out into the wilderness on the Jewish Day of Atonement, standing in a salt marsh near Mount Sodom.

12. Bedouins sitting by the northern shore of the Dead Sea in the late nineteenth century. In the background on the left is Rujum el-Bahr, an artificial breakwater constructed in antiquity. Today, Rujum el-Bahr stands on dry land approximately a kilometre north of the shoreline.

13. Turkish trenches on the shores of the Dead Sea, 1918. The area was conquered by British forces and their allies towards the end of the First World War.

14. Aerial photograph of the northern Dead Sea in the 1930s. The Jordan River flows into the lake with Palestine Potash Company evaporation pools on its eastern and western banks.

15. The traditional baptism site on the Jordan River, a few kilometres north of the Dead Sea. Visitors to Qasr al-Yahud on the western bank are just a few metres away from the border with the Kingdom of Jordan that runs through the middle of the river. In the background, one of the modern churches of Bethany Beyond the Jordan.

16. Part of the remains of the Byzantine-era complex of churches at Bethany Beyond the Jordan. The black marble steps leading to a bathing area at what must have been the river level at the time, before the course of the Jordan shifted, illustrate the site's association with baptism.

17. A rock formation on the cliffs east of the Dead Sea. Over the centuries, local guides have pointed to a number of features, including this one, as being the pillar of salt that was once Lot's wife.

18. Salt accumulates on the western edge of an evaporation pool south of the hotels of Ein Bokek.

19. Brown floodwater from the hills in the east spreading on the surface of the Dead Sea. Most of the rivers and streams that used to flow into the lake are diverted and no longer reach it. Flash floods in the winter help to slow down the falling water level.

20. The terraces left by the receding shoreline on the eastern side of the Dead Sea. The level of the lake drops by about a metre per year.

21. A view of the hills west of the Dead Sea from its eastern side. Wherever the water touches the shore, the whitened rocks, coated with salt, serve as a reminder: this is no ordinary lake.

22. One of the thousands of sinkholes created by the receding shoreline on the western side of the Dead Sea. Although walking alongside the sinkholes could be dangerous, they have become an unregulated tourist attraction.

23. The southern basin of the Dead Sea as captured from space by a CORONA satellite in the late 1960s. Some evaporation pools can be seen south of the lake, on the Israeli side, and a long dike marks the international border with Jordan.

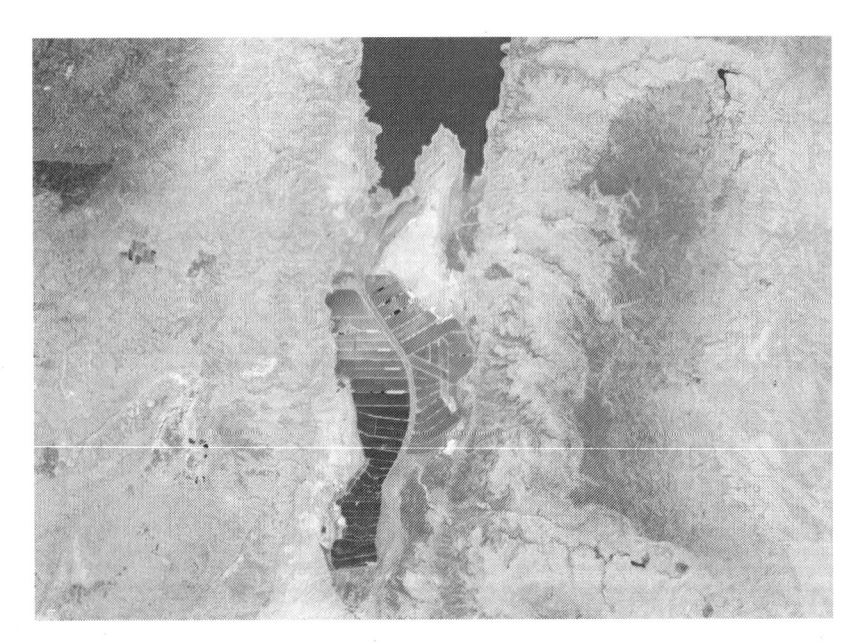

24. A contemporary Sentinel 2 satellite image of the southern basin of the Dead Sea. This part of the lake has been completely replaced by evaporation pools. The Lynch Strait has dried up, so water for the pools has to be pumped from the shrinking lake in the deeper northern basin.

CHAPTER FIVE

※

A SITE OF SCIENCE, EXPLORATION AND DIVERGING NORMS

DRIVING SOUTH ALONG Route 90 on the western side of the Dead Sea, one can easily be distracted by the lush vegetation of the Ein Feshkha nature reserve on the left and overlook the PEF rock standing above the road on the right. The rock received its name from a mark left in 1900 by two British explorers of the Palestine Exploration Fund who wanted to measure the Dead Sea's fluctuating level. The PEF explorers sought to use the rock near Ein Feshkha to scientifically record to what extent the level of the lake rose and fell on a yearly basis. The effort was led by Dr Ernest William Gurney Masterman, who, alongside his role in the Palestine office of the PEF, served as the Director of the Anglican Hospital in Jerusalem. Years later he explained that 'we had a mark cut upon the rock 14 feet above the then Dead Sea level – this height being chosen because it was a convenient one for dropping the yard measure to the surface of the sea'.[1] Today, one has to park one's car on the side of the road and climb to see the line they etched into the rock, which stands hundreds of metres away from the retreating Dead Sea shoreline.

The PEF rock is a reminder not only of how high the level of the lake was at the turn of the twentieth century, but also of the lengthy efforts of numerous scholars to study some of the unique attributes of the Dead Sea. While pilgrimage to holy sites near the lake and Bedouin trade caravans along its shores continued, the nineteenth century heralded an age of scientific exploration. European and North American explorers came to the Dead Sea imbued with the belief that they could unlock its secrets using the latest technology and scientific methods. At the same time, rival imperial powers sought to use geographic exploration in the Middle East to further their strategic interests. Let us examine how the body of knowledge surrounding the lake changed and developed during the nineteenth and early twentieth century, focusing on three lines of inquiry: the reason for the fluctuating level of the lake's shoreline, the depth of the Dead Sea in relation to the Mediterranean, and the study of the fauna by the lake's shore, as well as whether any creatures were capable of living inside it. Improved scientific understanding led to two conflicting ambitions as to how best to use the Dead Sea: constructing canals to bring sea water into the lake or exploiting its existing mineral resources. Meanwhile, the chief residents of the area, the Bedouins, benefited from the stream of visitors but were not consulted when plans to transform the lake were drawn up.

QUESTIONS OF LIFE AND DEPTH

The Dead Sea is a terminal lake in which its chief tributary, the Jordan River, 'is swallowed up and never seen again'.[2] Why did the level of the lake fluctuate? This is a question that travellers and scholars had long grappled with. In antiquity and in the medieval period, there were those who suggested that the Dead Sea must be connected underground with the Mediterranean. This supposed connection would explain why the level of the Dead Sea did not constantly rise, even though water was flowing into it while none appeared to pour out.

A SITE OF SCIENCE AND EXPLORATION

Perhaps water was or had been flowing out of the Dead Sea on its southern edge? In 1838 the French Société de Géographie tasked the Beirut-based explorer and topographer Jules de Bertou with examining whether water had once been able to flow between the Dead Sea and the Red Sea. Along with local guides and guards, he ventured from Jerusalem through Hebron down to the south-western shores of the Dead Sea. Passing by Mount Sodom, he continued southwards along Wadi Arabah, making measurements on his way. After about three days of travel from the southern tip of the lake, de Bertou and his companions reached a relatively high point in the valley separating the direction of ravines: some flowed north to the Dead Sea while others discharged south towards the Red Sea. With this watershed in mind, he concluded that 'in the present state of things, the river Jordan never could have flowed' from the Dead Sea to Aqaba on the shore of the Red Sea.[3]

While investigating the hydrology of the Dead Sea basin, de Bertou and other European explorers realised that the lake and its surrounding area were considerably lower than the Mediterranean. The method de Bertou used to work out elevation required that he boil water repeatedly.[4] At normal sea level, water boils at 100°C. If one were on top of a mountain, the boiling point of water would be lower because of the reduced atmospheric pressure. However, near the Dead Sea, atmospheric pressure is greater and thus water boils at a higher temperature. That water boiled at different temperatures, depending on the altitude, was already known in the early nineteenth century. However, it took some time to develop thermometers that could measure temperatures accurately. Two British explorers, George Henry Moore and William G. Beek, were the first to test this method in the Levant, in 1837, though their equipment was not sufficiently precise to work out the exact altitude of the Dead Sea. Using 'common rain water' in a special tin intended for this purpose, they received the same result when boiling water in Jaffa and Beirut, both along the Mediterranean coast. However, next to the Dead Sea, the temperature in which water began to boil was indeed higher. They concluded that there was a 'remarkable feature in the level of the sea, as from several observations upon the temperature of boiling water, it appears to be considerably lower than the ocean'.[5]

Jules de Bertou was suspicious of Moore and Beek's findings. He therefore conducted a number of measurements over the course of a few years and, eventually, conceded that the Dead Sea was lower than normal sea level. He first estimated that the level of the Dead Sea was 422.7 metres lower than the Mediterranean, and later amended his estimate to a difference of 406.13 metres. The next major contribution to this scientific debate came from an American expedition led by the naval officer William Francis Lynch in 1848. Lynch and his companions mapped the Jordan River from Lake Hula in the north to the Sea of Galilee and from there all the way south to the Dead Sea. Based on barometrical measurements, they calculated that the latter lake was 1,316.7 feet (401.33 metres) lower than the Mediterranean, the most accurate calculation thus far.[6] The strait connecting the Dead Sea's northern and southern basin was later named after Lynch.

While the PEF was being formed in London in 1865, British surveyors undertook what they called 'levelling from the Mediterranean to the Dead Sea'. Rather than measuring the temperature of boiling water, the method they used calculated the difference in height between fixed points on the ground. Colonel Sir Henry James and his team cut fifty-five survey benchmarks on fixed rocks at intervals between Jaffa on the Mediterranean coast, Damascus Gate in Jerusalem and the shore of the Dead Sea. The height they recorded at the shore of the lake was -1,292.135 feet (393.84 metres) below the level of the Mediterranean.[7] This levelling exercise set the scene for a far more comprehensive endeavour to map the southern Levant. As we have seen, in the following decade, the PEF carried out a detailed survey of Palestine, partly to satisfy the interests of its members but partly also because of British imperial strategic needs. Although the country was still under Ottoman rule, mapping the Jordan Valley in particular was seen as a means to better protect British interests in the Suez Canal and the route to India against potential threats from France and Russia. As the British War Office confirmed in 1877, a good map of the country had 'enormous military value'.[8]

With the question surrounding the depth of the Dead Sea below normal sea level settled, attention returned to recording and accounting for the lake's rising and falling shoreline. Before the PEF line was put in place, the main tell-tale sign for the fluctuating level of the Dead Sea was the artificial mound known as Rujum el-Bahr near the northern edge of the lake. As we saw in Chapter 3, in the late Hellenistic and early Roman period, the mound had a built structure on it and served as an artificial breakwater and anchorage point in an area exposed to winds and where the gently sloping lakebed consists of fine-grained sediments with no local boulders to which boats could be fastened.

Over the years, several travellers and pilgrims commented on the mound but described it in different ways. For instance, the American naval officer Lynch, who sailed past Rujum el-Bahr in 1848, marked it on his map as a peninsula. In 1869, the Italian diplomat Paolo Bajnotti reported that, ten years earlier, the ruins on the mound had been about a metre underwater, whereas these had since then been revealed by the level of the lake dropping. British officers Claude Conder and Horatio Herbert Kitchener, who inspected the area in 1873–1874 as part of the PEF's Survey of Western Palestine, noted correctly that it must have served as 'an artificial pier constructed at some period when boats were used on the lake'.[9] The map they produced showed the mound as barely linked to the shore through a very narrow isthmus. The British solicitor and travel writer John Gray Hill observed that the mound used to be an island one could swim to in 1890, but by 1897, it had become completely submerged. The appearance and disappearance of Rujum el-Bahr depended on the changing level of the lake. We now know that when the water level was higher than -392.77 metres it was fully submerged. At times when the level of the lake was between -392.77 and 399 metres, it appeared as an island. The Rujum was a peninsula linked to the mainland by a low-lying isthmus at water levels of between -399 and -400 metres. Finally, when the lake level dropped below -400 metres, the site became exposed on the dry lakebed. Today, Rujum el-Bahr stands on land approximately a kilometre north of the shoreline.[10]

There were two schools of thought when it came to accounting for the fluctuating level of the Dead Sea. One school of thought was represented by Charles F. Tyrwhitt Drake, who had joined the PEF's Survey of Western Palestine as a volunteer in the 1870s. Drake noted that in the first half of the nineteenth century, it had been possible to ford the lake where its two basins meet, near the Lisan Peninsula. However, by 1874, it was no longer possible to walk across from Lisan in the east to the western side of the lake owing to the depth of the water. Noting also that the causeway connecting the mainland with Rujum el-Bahr had become submerged, he argued that the bottom of the Dead Sea was subsiding. A few decades later, John Gray Hill, who drew on Drake's writing, pointed out that changes in the lake's level did not correspond fully to the prevalence of rain. Noting that the level of the Dead Sea had been rising in the late nineteenth century, he speculated that 'there is some volcanic action at work raising the bed of the lake'.[11]

The second school of thought had a simpler explanation for the changing level of the water. One of its proponents was the orientalist and Cambridge professor Edward Henry Palmer. He pointed out that, despite the great volume that flows into the Dead Sea, and even though 'there is no apparent, or, indeed,

A SITE OF SCIENCE AND EXPLORATION

conceivable outlet, the immense evaporation which takes place' was 'sufficient to maintain the level of the lake'.[12]

Eventually, it was the routine measurements using the mark on the PEF rock which laid the speculations to rest. Nearly every year from 1900 till 1913, Masterman and his PEF colleagues dutifully measured and recorded the high level of the lake in March, at the end of the rainy season, and its low level in November, at the end of the region's dry season. Some years were drier (1904), some were wetter (1911), but the measurements in tandem with precipitation figures from across the country enabled Masterman to conclude that 'rainfall has much to do with the varying level', alongside the height of temperatures during the summer and 'the effect of a series of dry seasons in the diminution of the springs'. Masterman was also aware of the influence of the extent of snowfall upon Mount Hermon at the northern edge of the Jordan River catchment.[13] In short, the fluctuating level of the Dead Sea was the result of the interplay between rainfall and evaporation.

NOT QUITE DEAD

Aristotle and several other classical and medieval authors had noted that no creatures could live in the Dead Sea. Nonetheless, the search for organisms that could survive the lake's saline conditions continued through the centuries. For instance, Henry Maundrell, who visited the lake in 1697, questioned whether no fish could endure its deadly waters. He reported 'having observed amongst the pebbles on the shore two or three shells of fish resembling oyster-shells'. He noted that this shore was about two hours' travel from the mouth of the Jordan, therefore making it unlikely that these had arrived in the Dead Sea from the river.[14] The Swedish botanist Fredrik Hasselquist, a student of the great Carl Linnaeus, was inspired by his teacher to study the flora and fauna of the Holy Land. He visited the Dead Sea area in April

1751 and was very glad to have his herbarium with him: it served as a pillow when he camped near Jericho, while his companions had to lie on the bare earth. Hasselquist identified several plants and birds near Jericho and along the Jordan River. His stay by the Dead Sea itself was brief, but, like Maundrell, he reported that shellfish were common on the shore.[15] This assertion was countered by Ulrich Jasper Seetzen following his exploration of the area in 1806. Seetzen took pride in spending much more time by the lake than Hasselquist had. He pointed out that he could not find even one mussel, and that any shells on the shore belonged to land snails.[16]

European travellers of the late eighteenth and early nineteenth century began to collect samples of Dead Sea water and bring them back to Europe for analysis. These found no traces of life. One of those who took a sample for analysis was Richard Robert Madden. This Irish physician visited the Dead Sea in 1824, disguised as a Bedouin. Madden sought to prove there was no life in the Dead Sea by trying to fish in the lake for two hours, catching only bitumen.[17]

Perhaps the best-known nineteenth-century naturalist to explore the Dead Sea area was the aforementioned clergyman Henry Baker Tristram. He spent ten months exploring Palestine's natural history and geology with a small group of friends in 1863–1864. Summarising his travels, he emphasised that 'Our attention was particularly directed to the basin of the Dead Sea, and to the districts east of the Jordan, as being those least accessible to travellers, and of which our knowledge was least complete.'[18]

Tristram was thoroughly impressed by the natural beauty of the Jericho oasis: 'in zoology Jericho surpassed our most sanguine expectations. It added twenty-five species to our list of birds collected in the tour, and nearly every one of them rare and valuable kinds.' Among the many birds in the trees and shrubs of the oasis, he spotted large numbers of *bulbul* (*Pycnonotus xanthopygos*),

Palestine sunbirds (*Cinnyris osea*) and a species of starling with black plumage and orange patches on the outer wing that came to be known as Tristram's grackle. The reverend also observed scorpions 'under every stone' and a fine though 'decidedly poisonous' snake, the cerastes of the Dead Sea.[19] By the banks of the Jordan, he spotted traces of wild boar, hyena and jackal. In one of the gorges leading to the Dead Sea, he even found the lair of a leopard. It is, however, doubtful whether present-day zoologists would approve of Tristram's method of collection. He and his party used to shoot down their specimens before examining them. One of Tristram's Bedouin guards could not understand why they were shooting animals 'neither to eat nor to sell'. The guard thought 'we had some spells, by which we should restore all the birds to life when we got into our own country'. Why these 'lords' should use witchcraft to bring serpents into their country was beyond him.[20]

By the northern shore of the lake, Tristram found 'many blanched and almost calcined specimens of Jordan shells' along with 'quantities of very small dead fish'. As the visit took place in January, he judged that the fish were brought down the Jordan River by floods and killed by the salty water of the Dead Sea. The travellers observed several gulls and shot a kingfisher, with Tristram remarking on 'how utterly absurd are the stories about the sea being destitute of birds'. With regard to life in the lake itself, however, he was adamant: 'it is quite certain that no form of either vertebrate or molluscous life can exist for more than a very short time in the sea itself, and that all that enter it are almost immediately poisoned and salted down.'[21]

That the water of the Dead Sea must not be ingested is clear from the story of Christopher Costigan, a theology student from Maynooth College, Ireland. Like a number of explorers after him, Costigan had a small boat transported overland from the Mediterranean to the Sea of Galilee. From there he sailed down the Jordan River all the way to the Dead Sea. His aim was to explore the lake's shores, measure the depth of its water and account for its buoyancy. This journey took place in late August 1835, a time of immense heat. Costigan, who was already exhausted, reportedly made his coffee by boiling Dead Sea water. After eight days of sailing and exploration, he became very ill. He was brought to Jerusalem, where he died in early September. While the exact cause of his death cannot be determined, one modern-day nephrologist, Professor Michael Friedlaender, pointed out that the water of the lake contains potentially toxic levels of magnesium and calcium. Ingesting even a small amount of Dead Sea water is sufficient to cause extreme hypercalcemia and hypermagnesemia, which could prove fatal.[22]

With the chemical composition of the lake's water in mind, the results of an analysis carried out on samples of Dead Sea mud by the French scientist Louis Lortet in 1892 proved very surprising. Lortet, the Dean of the Faculty of Medicine of Lyon, had taken part in an expedition to the Dead Sea in the 1870s, but then he found that the lake does not contain 'any living organism, plant or animal'. He speculated whether Dead Sea water could be used as some sort of aseptic liquid. However, years later, when he was given samples of mud from the lakebed which he divided into many tubes and flasks, he was astonished to observe after forty-eight hours that these contained two microorganisms, especially in their deeper parts: *Clostridium tetani* (a common soil bacterium that causes tetanus) and *Clostridium perfringens* (an anaerobic bacterium that is associated with traumatic gas gangrene), which can grow where there is no oxygen. He concluded that 'certain pathogenic microbes may resist ... prolonged contact with large masses of water, even whilst they contain, in considerable quantities, salts injurious to every other organism, animal or vegetable'.[23] Alas, Lortet's discovery went largely unnoticed.

It wasn't until the 1930s and 1940s that the microbiologist Benjamin Elazari Volcani (Wilkansky) managed definitively to

A SITE OF SCIENCE AND EXPLORATION

identify a few types of halophilic microorganism that can tolerate the extreme conditions of the Dead Sea. The lake turned out to be not entirely dead after all. First, in 1936, the then young graduate student from the Hebrew University of Jerusalem took samples of water from the lake at various depths while staying 3 to 4 kilometres away from the estuary of the Jordan River. Upon analysis, the water contained three microorganisms. Further research in the following years enabled Volcani to isolate a number of red halophilic archaea – single-celled prokaryotic microorganisms that lack a cell nucleus – as well as a number of halophilic bacteria that thrive in salty environments. He was also able to identify algae species. When his research was conducted, the green algae *Dunaliella* was prevalent in the upper water layer, but as we will see later on, this has changed since then with the increasing salinity of the lake. Volcani pioneered using cores of sediments from the bottom of the Dead Sea, a process that gave an indication of the layering that future scientists would go on to study. 'On dissecting the profile longitudinally, a beautiful "spectrum of layers", of different colours – black, dark blue, grey, brown, and white – was revealed', he noted in 1943. The study of sediments from the bottom of the lake enabled him to identify a number of anaerobic bacteria. Volcani's achievements in the field of halophilic bacteria and algae did not go unrecognised. The bacterial genus *Volcaniella* and the *Haloferax volcanii*, a type of archaea, were named after him.[24]

'DYNAMICAL GEOLOGY'

The Dead Sea is situated in a rift created by plate tectonics. The theory of 'continental drift' was first put forward in the 1910s by the German meteorologist and geophysicist Alfred Wegener. The mechanics of plate tectonics were not fully understood until the Cold War. Seismometers that monitored nuclear experiments showed that earthquakes and volcanic eruptions occurred in the vicinity of distinct belts around the world. Echo-sounding technology and geophysical exploration of the ocean floor, developed by the US Navy to track Soviet submarines, paved the way for the discovery of ocean ridges and underwater fracture zones.[25] However, earlier exploration in the Dead Sea area in the second half of the nineteenth century enabled geologists and geographers to make observations about movements and fissures in the earth's crust many decades before plate tectonics became widely accepted.

The former Royal Navy officer and Fellow of the British Royal Geographic Society Captain William Allen (1792–1864), who visited the Dead Sea in 1850, was one of the first to suggest that 'the depression, in which we find this inland sea, was originally formed by some subterranean movement, in common with the general production of mountains and valleys on the earth's surface'. He also observed that the formation of the lake 'was only a part of a very large operation, which included also the Gulf of Akabah and the Red Sea; since they are in the same line of action, and in some respects are similar'.[26] As we will see later on, Allen's plan to alter the landscape of the Dead Sea was not taken forward. As a result, his ideas about how the lake was formed did not attract much attention either.

A few years later, the Reverend Tristram also offered some initial thoughts on how the Jordan Valley was first formed after he and his companions descended from Jerusalem to Jericho in late December 1863. He observed that 'The lower strata appeared, as a general rule, to dip evenly to the east-ward, as if the Ghor (or Jordan valley) had been, after the secondary geological period, gently and gradually let down.'[27] However, Tristram still believed that 'only the Jordan, or that ancient tongue of the Red Sea which it represents, could have formed the Ghor, and especially the strange gravel hills around us, and that the Ghor is chiefly a fissure of erosion'.[28] In other words, Tristram believed that although a depression induced by volcanic activity may have contributed to its shape, the Jordan Valley was formed, essentially, by the flow of water.

The British geologist Wilfrid H. Hudleston, who was acquainted with Tristram and his work, was much bolder when he published *The Geology of Palestine* some two decades later. Hudleston speculated that a 'fault must be concealed under the Dead Sea itself', forming part of a larger system that stretched from Lebanon in the north all the way down to the Red Sea. He recognised the broader importance of understanding how the Jordan Valley was formed, as it raises 'the possibility of a sinking-in of the earth's crust at intervals along certain lines of fissure – an explanation so distasteful to most stratigraphists'. He was not quite ready to embrace this interpretation, speculating instead that the valley may have once been 'a hole in the ocean floor', and that the floor of the ocean had since then risen. Based on the work of the French geologist Louis Lartet from the 1860s, Hudleston recognised that there once was a larger 'Jordanic lake' – what we now call Lake Lisan – that at one point covered large parts of Wadi Arabah, the Dead Sea itself and the Jordan Valley further to the north. Hudleston suggested, correctly, that there was once a connection between the inland lake and the Mediterranean coast through the Jezreel Valley.[29]

It was a fellow British geologist, Edward Hull, who came closest to describing plate tectonics. Hull surveyed southern Palestine and Transjordan for the PEF and published his contribution to the Survey of Western Palestine in 1886. Here is what he had to say about the process he called 'dynamical geology': 'we have to deal with terrestrial forces acting within the earth's crust on the one hand, and the various agents of denudation or erosion which have operated from without, on the other.' He notes that, until the close of the Eocene period, North Africa, the Arabian Peninsula and Syria were at the bottom of the ocean. The floor of this ocean accumulated 'successive beds, chiefly of limestone, [which] had been laid down during the Cretaceous and Eocene periods, until they had reached a thickness of several thousand feet'. Then, the ocean floor was lifted 'owing to movements which elevated out of the waters the present land areas, and depressed to still greater depths other portions of the same general sea-bed'.[30]

While studying the formation of the Jordan Valley and the Arabah, he observed 'That this deep depression is the direct result of a "fault", or fissure, of the crust, accompanied by a displacement of strata, the relations of the formations on opposite sides leave no room for doubt.' Hull even noticed, much like subsequent geologists in the twentieth century, that 'there is a general dissimilarity in the geological structure of the opposite sides of the Jordan-Arabah Valley', and that 'formations occupying the eastern side are older than those of the western'. Considering he conducted his survey in the nineteenth century, it is remarkable that he was able to explain the differences between the eastern and western sides of the valley using a terminology that was very much ahead of its time: 'This general dissimilarity in the stratification indicates a displacement along a line of fault, to an extent of several thousand feet in some cases.'[31]

In general, Hull anticipated many of the ideas put forward by present-day geologists such as Aharon Horowitz, discussed in the first chapter of this book. Hull argued that there is one central and continuous line of dislocation along the Jordan Valley depression but that at times, in the Arabah, 'the actual fault or fracture is not often visible'. Alongside the main north–south fault there were 'numerous minor fractures'. Hull understood that 'whatever the cause of the terrestrial disturbances along the special line of the Jordan-Arabah depression, it is sufficiently clear that the line was an axis of disturbance for the whole region'. He was also very perceptive about how these geological changes may have affected the salinity of the lakes that preceded the Dead Sea:

A SITE OF SCIENCE AND EXPLORATION

> The increase of saltiness in the waters of the Dead Sea has probably been very slow, and dates back from its earliest condition, when its waters stretched for a distance of about 200 miles from north to south. While the uprising of the land, and the sinking down of the Jordan-Arabah depression were in progress during the Miocene period, some of the waters of the outer ocean, themselves salt[y], were probably enclosed and retained; but from the occurrence of the shells in the marls in the Arabah Valley ... it would appear that, when the waters of the great inland lake were at their maximum elevation, they were sufficiently fresh to allow of the presence of molluscous life.[32]

As his observations on the Dead Sea area show, Edward Hull certainly deserves a place of honour among the precursors of the plate tectonic revolution.

By the early twentieth century, the idea that the rift in which the Jordan Valley and Dead Sea lay was formed as a result of subterranean geological activity was certainly taking hold. The physician Masterman noted in 1905 that 'This depth, along the line of the great "fault" – the line of shifting of the earth's crust, the cause of this great rift – is exceptional.'[33] A few years later, the American geographer Ellsworth Huntington observed that 'a river whose valley has not been made by the action of water, but by the movement of the earth's crust, is one of the rarest of phenomena'. He describes bathing in a lagoon near the northwestern shore of the lake that was 'warmed up from the heated depths along one of the many fissures characteristic of the faults whereby the plateaus have been separated'.[34] It is not surprising that when Wegener came to articulate his interpretation of the mechanics of how continents and oceans were formed, one of the examples he drew on was the 'large fault system which can be traced northwards through the Red Sea, the Gulf of Aqaba and the Jordan Valley to the edge of the Taurus fold range'.[35]

ALTERATION

With the growing understanding of the unique topography of the Dead Sea, many began to suggest that radical changes could be made to the landscape so as to better serve changing human needs. One recent example is an idea put forward by the French-Spanish author Tomas Pueyo Brochard. In May 2023, he posted the following suggestion for the Dead Sea area on Twitter (before it was renamed X):

> It's 200-600m below sea level and there's not much there: some salt harvesting, a bit of tourism ... It's mostly a desert. You could bring water from the sea and flood it. This would mean:
> - Electricity generation, given the huge depression
> - Which could be used for desalination. Fresh water. Irrigation. More life.
> - The area would get more moisture, with more plants & animals
> - More tourism[36]

The idea to flood the Dead Sea is not particularly new. From the mid-nineteenth century onwards, improved cartography and technological advancements in engineering and explosives paved the way for the construction of massive canal projects. In the case of the Suez Canal, the Corinth Canal and the Panama Canal, these complex and expensive projects came to fruition. This wasn't the case with ideas involving the Dead Sea.

One of the early proponents of altering the landscape of the Dead Sea area was Captain William Allen. Having visited the southern Levant in 1849–1850, Allen first sent to Lord Palmerston, then a key minister in the British government, a 'Plan for a ship canal to India by the Dead Sea'.[37] Allen subsequently elaborated his ideas in a book, *The Dead Sea: A New Route to India*, published in 1855. Aware of the concession given by the Ottoman Sultan to the French to construct a canal from the Mediterranean to the Red Sea across the Isthmus of Suez, he nonetheless put forward an alternative plan. He argued that a canal could be dug from Kaiffa (Haifa) on the Mediterranean coast, allowing sea water to run along the Jezreel Valley so that most of the Jordan Valley would be flooded. The difference in height between the Mediterranean and the Jordan Valley, he believed, would make construction efforts easier, as it would create a current that would carry off all the earth 'previously loosened by blasting'. Another canal would then be dug from the very much enlarged Dead Sea to Aqaba on the Red Sea, thereby creating a shorter 'means of communication with our East Indian possessions'. With the Mediterranean and Red Sea connected, ships would no longer need to take the longer route to India around Africa.[38]

Allen's plan was never pursued, and not only because the Suez Canal was inaugurated in 1869. The geologist Hudleston was very dismissive of the logic behind Allen's suggestion. 'A more childish idea was probably never put into print', he wrote in 1885. Hudleston pointed out that any attempt to connect a flooded Jordan Valley with the Red Sea would be completely impractical because of the difficulties of cutting through the higher elevation of Wadi Arabah. He also put forward moral reasons for opposing the plan. In words that could have been repeated by environmental activists and conservationists today, Hudleston pointed out that by flooding the Jordan Valley with Mediterranean water 'the remarkable scenery and physical peculiarities of the Dead Sea, with all its ramifying valleys, would be forever destroyed, and the most curious feature on the surface of the earth obliterated'.[39]

While the idea to sail ships from the Mediterranean through the Dead Sea down to the Red Sea was abandoned, the ambition to alter the landscape to further human needs remained. Such ambitions were eagerly picked up by the leaders of the Zionist movement: Jews, predominantly in Europe, who from the late nineteenth century onwards sought to create a national home for the Jewish people in Ottoman Palestine, which they saw as their historical homeland. In the 1890s, Theodor Herzl, one of the founding fathers of the Zionist movement and its first major leader, received a number of proposals that advocated bringing water from the Mediterranean to the Dead Sea through a canal or a tunnel in order to generate hydroelectricity. One of these was sent to Herzl by the Swiss engineer Abraham Max Bourcart in 1899. It included a sketch of a canal from the Mediterranean to the Jordan Valley through the Jezreel Valley. Herzl was inspired by the suggestions he received. He included a description of the modernised Dead Sea, benefiting from hydroelectricity, in his utopian novel, *Altneuland* (Old-New Land), published in 1902.[40]

The protagonists of *Altneuland* return from a remote island, where they had spent twenty years, and pass through Palestine. While they had been away, the country had been completely transformed by the Zionists. This thriving old-new land is powered by hydroelectricity produced at the Dead Sea. Herzl describes a canal, 10 metres wide and 3 metres deep, that brings water from the Mediterranean, which then drops through a tunnel to a series of power stations. The protagonists visit one of these stations on the north-western edge of the lake, near Jericho, and are taken aback by the roaring sound of the water. One of the guests inquires how the level of the lake remains steady, with so much additional water coming in. A local professor assures him that as much water is taken away from the Dead Sea as reaches it through the canal. Great quantities of freshwater – presumably from the Jordan – are diverted into reservoirs and used for irrigation 'in areas where water is as necessary as it is superfluous here'.[41]

Herzl died in 1904, and a hydroelectric plant on the north-western edge of the Dead Sea has not been built. However, the idea to construct artificial canals leading to the lake would reappear in various guises in the following decades. Furthermore, the suggestion that freshwater could be diverted away from the Dead Sea would be implemented in the second half of the twentieth century, as we will see later on.

THE BEDOUIN LAKE

As well as being a time of intense scientific exploration, the nineteenth century saw various depictions of the Dead Sea in European literature and art, which were dominated by religious images and the role the area had played in historical events. The Scottish author Walter Scott began the novel *The Talisman* (1825) with a description of a Crusader knight standing on the shore of the lake and contemplating 'the fearful catastrophe, which had converted into an arid and dismal wilderness the fair and fertile valley of Siddim'. Once well-watered, all the 'warlike pilgrim' could see was 'a parched and blighted waste, condemned to eternal sterility'. The Crusader recalls that 'beneath these sluggish waves lay the once proud cities of the plain', whereas now the odour of bitumen and sulphur was keeping even the birds away from the lake. Standing in his unwieldy defensive armour, the northern Crusader seemed to defy the climate and country to which he had come.[42] While Scott never visited the Dead Sea, the English pre-Raphaelite artist William Holman Hunt did. In 1854 he began working on *The Scapegoat*, a painting inspired by the Book of Leviticus, showing a goat that had been sent out into the wilderness on the Jewish Day of Atonement with a piece of cloth on its head, symbolising the sins of the congregation. Hunt has the parched goat standing in a salt marsh near Mount Sodom – the site his contemporary, the French explorer Louis Félicien de Saulcy, had identified with biblical Sodom. The skeletons of other animals, part of the lake and the hills of Edom can be seen in the background. It seems that an early version of the painting, with a rainbow behind the goat, did not convey properly the desolate image that Hunt wished to convey. Hence, the final version of the painting, on display at the Lady Lever Gallery in Liverpool, omitted the rainbow and further emphasised the lifeless setting.[43] The Dead Sea also served as the backdrop for the short story 'Hérodias', published in 1877 by the French author Gustave Flaubert. The story centres on the beheading of John the Baptist at the fortress of Machaerus near the eastern shore of the lake. Flaubert depicts the fortress as having richly carved archways, arched windows and luxurious gardens. It is surrounded by deep valleys and offers a view of the rugged but beautifully lit hills of Judea on the opposite shore of the Dead Sea. The story likely inspired Oscar Wilde's play *Salome*, written a few years later.

While depictions by nineteenth-century European authors and artists harked back to antiquity and the medieval period, throughout the Ottoman period the Dead Sea itself remained a Bedouin lake. A nomadic people whose society was structured around oral traditions, the Bedouins who lived around the Dead Sea in this period did not leave behind written texts for historians to use. The sources we do have were written by non-Bedouins, and these often exhibit a mixture of awe, suspicion and fear when describing these residents of the desert. Most European and North American explorers believed in the superiority of their own culture and held condescending views towards the Bedouins. However, although the sources we have are far from ideal, some observations can be gleaned from them about how the Bedouins managed the area around the Dead Sea.

As we have seen, the presence of Bedouins in the Dead Sea area is mentioned in Crusader sources from the twelfth century. However, at some point after the thirteenth century, new Bedouin groups moved into the southern Levant. Over time, different tribes inhabited and controlled different areas, at times displacing one another. The German explorer Ulrich Jasper Seetzen, who toured the southern Levant in 1806–1807, compiled a list of the Bedouin tribes he encountered. All the major groups he noted appear consistently in subsequent accounts.[44] These tribes engaged in herding, the transportation of goods and people, and, in some cases, small-scale arable agriculture.

When the PEF team conducting the Survey of Western Palestine completed its work in the Jordan Valley in May 1874, Drake listed which tribe saw which area as its own, as well as estimating their size. The list included the largest tribe in the vicinity of the lake, the Ta'amirah (or Ta'amireh), with about 1,000 men and some 400 tents, who dwelled in the desert north-west of the Dead Sea, up to Bethlehem. The Jahalin (or Jehalin) lived further to the south, near Hebron, and included some 150 men.[45] Another group was the Ghawarna (also known as Ghawarni or Ghawarineh), the Arabs of the Ghor, the Arabic name for the Jordan Valley. A number of travellers mention this latter group as dwelling in the plain of Jericho. For instance, John Wilson of the Royal Asiatic Society wrote in 1847 of their encampment of about fifty tents, a little west of Jericho. He notes they have dark skin and 'are much more demoralized than their brethren of the desert'.[46] Either because of their dark complexion or because of their association with historic slavery, the Ghawarna who lived near the southern edge of the Dead Sea were viewed as socially inferior by other tribes in the Karak region.[47]

Foreigners often felt compelled to hire groups of Bedouins to ensure their protection. According to Lynch, 10 miles east of Jerusalem 'the tribes roam uncontrolled, and rob and murder with impunity'.[48] Allen, who visited the Dead Sea a short while after Lynch, observed that 'Security is obtained by the traveller taking certain of the robbers into his service'. Not without humour, he described the negotiations preceding any trip to the Dead Sea area, which:

> involved long discussions and altercations with the sheikh of the Arabs, who is deputed to be the receiver of the toll or tax which they pretend to have the right of levying on all persons entering, not their country, but the tracts abandoned by civilization, and overrun by them ... a person once placed under their protection is in security in the desert – guarded by professional robbers, even among rival tribes, who make it a point of honour to respect such protection[49]

He later conceded that 'Our sheikh' was 'one of the noblest-looking fellows I ever saw'.

In the 1860s there were clashes between the Ottoman authorities and Bedouin tribes near the Dead Sea – the same clashes that made it so difficult to secure the Moabite Stone. That the area was restive during those years is apparent in the accounts of travellers. From his Jahalin guide and guard, Tristram learned of raids and counter-raids between tribes west and east of the Dead Sea. The raids resulted in camels being taken and some men getting wounded. 'We saw here a perfect specimen of Arab warfare, and of the state of the country', he remarked. However, he felt compelled to add that 'their battles are seldom bloody, and the vanquished party usually emigrates at once'. In general, Tristram was reassured by the experience of previous travellers, like the Frenchman de Saulcy who 'had had no difficulties with the Arabs after backshish had been settled, excepting on one occasion, when they had exchanged a few harmless pistol-shots'.[50]

While initially Tristram may have had misgivings about how the Bedouins 'levy a blackmail', he later realised that the stories of unrest and fighting were not unfounded. When he and his companions stood next to the banks of the Jordan River north of the Dead Sea, the guards 'insisted upon the horses being at once withdrawn out of range from the river-bank, lest, if detected, they should afford target practice to their neighbours, the tribes on the other side, with whom they informed us they were at war'.[51] The danger became more palatable on the southern edge of the lake. When they reached the village of Safieh (Ghor es-Safi), they saw clear signs of violence: smoke was still rising from huts that had been burned, and some dead bodies still lay where they had been killed. Tristram was shocked by the response of his guards: 'With the true Bedouin instinct, they were plundering and searching for loot in all directions.'[52] Along with wheat and barley, lots of indigo was stored in the village, which had been attacked a day or two earlier as part of a raid or a war between two tribes. Tristram and his party were forced to abandon their plan to proceed north towards the Lisan Peninsula for fear the raiders would return. Instead, they headed back to the safer south-western side of the Dead Sea.

A few years later, Mark Twain described clashes between Bedouin tribes on the one hand and Ottoman garrison troops and residents of the Jericho area on the other. While in Jerusalem, the American author had heard alarming news:

> The lawless Bedouins in the Valley of the Jordan and the deserts down by the Dead Sea were up in arms, and were going to destroy all comers. They had had a battle with a troop of Turkish cavalry and defeated them; several men killed. They had shut up the inhabitants of a village and a Turkish garrison in an old fort near Jericho, and were besieging them. They had marched upon a camp of our excursionists by the Jordan, and the pilgrims only saved their lives by stealing away and flying to Jerusalem under whip and spur in the darkness of the night.[53]

Twain later realised that the stories were exaggerated and, although he and his party were quite fearful, their visit to Jericho and the banks of the Jordan was uneventful. His judgment was similar to the one Allen had expressed a few years earlier: 'The nuisance of an Arab guard is one which is created by the Sheiks and the Bedouins together, for mutual profit.'

By the 1870s, things seem to have calmed down. Drake of the PEF's Survey of Western Palestine reported that 'From all the Arabs and Fellahin [peasants] in the Ghor we experienced nothing but civility'.[54] The British author and diplomat Laurence Oliphant had very little patience for the armed guard he was supposed to travel with. Although his route from As-Salt, east of the Jordan River, to Jericho was reported to be unsafe on account of Ghawarni Arabs, he was later very proud of himself for travelling without protection and having saved a lot of money.[55]

Like his predecessors, the American geographer Huntington heard numerous tales of Bedouins attacking foreign visitors, so much so that when visiting the Dead Sea area, he and his companions barely slept. 'It is hard to tell whether the danger was real or imaginary', he thought in retrospect. But Huntington showed more understanding than most for why the Bedouins acted as they did. A firm believer in how particular environments shaped different societies, Huntington pointed out that, if the spring season has been dry, the Bedouins do not have sufficient supplies of grass for their herds. Hence, 'according to their moral code which their environment has fostered, there is no reason why a man should not rob if he sees men of another race or tribe living in plenty while he suffers want'.[56]

Despite their biases, by reading against the grain one can nonetheless use the accounts left behind by European and North American explorers to identify a number of salient characteristics of how the Bedouins ran the Dead Sea area. First, the need for a Bedouin escort for those travelling in the vicinity of the lake was recognised by foreign consuls stationed in the Levant, and the latter were often involved in helping travellers negotiate the costs of securing an armed guard. It is possible that Ottoman authorities viewed the provision of an armed guard as a legitimate way

A SITE OF SCIENCE AND EXPLORATION

for the Bedouins to eke out a living while also providing reasonable assurance that foreign nationals would not be harmed.

Second, the Bedouins themselves respected the geographic limit of each tribe. Tristram, for instance, received the protection of one tribe for the Jericho area and another for the southern edge of the Dead Sea. Guides and guards became apprehensive when tribal boundaries were traversed. While travellers may have viewed such apprehensions as a mere performance, the Bedouins nonetheless had a clear sense of the informal geographic division of the land.

Third, once granted, the duty to protect foreign travellers was taken seriously. Tristram, for example, conceded that 'Their terms were not high for the country; and well and faithfully did Sheikh Mohammed and his men serve us during the whole time we were under their protection.'[57] Finally, travelling without adequate protection could be risky. The American missionary Archibald Forder, who worked at al-Karak around the turn of the twentieth century, once travelled with only two local escorts east of the Dead Sea. On their way they encountered 'a band of wild fellows that far outnumbered us, and helped themselves to such things as they fancied as they leisurely turned over our belongings, leaving us lighter than they found us, and me with a few bruises from blows of heavy clubs because I objected to their behaviour'.[58]

A properly negotiated armed guard of Bedouins not only served as deterrence but also usually commanded the respect of other tribes. Those foreign travellers who haggled and eventually secured Bedouin protection did not share Forder's unpleasant experience. It's not that the Bedouin-dominated Dead Sea area was lawless; it simply followed a different set of rules.

THE MOST MISERABLE VILLAGE IN PALESTINE

If the explorers' orientalist depiction of the Bedouins seems unkind at times, these still pale in comparison to the scorn with which the former described the Arab *fellahin* or peasants who lived by the Dead Sea. Only two small villages were recorded near the lake by nineteenth-century travellers: Ariha, near the site of ancient Jericho, and Ghor es-Safi (Safye or Safieh in some sources), near ancient Zoar. We know less about the latter, as far fewer foreign travellers reached it. The village itself seemed to Tristram very similar to Jericho – both were set in an oasis, and both were composed of 'wattled huts'.[59]

Most travellers only ventured as far as Jericho, so we know a little more about it. Back in 1751, the Swedish botanist Hasselquist lamented that the area around Jericho was almost entirely uncultivated, save for some barley the Arabs had sown for their horses. There was no building there except the walls of an old house, which the monks called the House of Zacchaeus.[60] Ariha must have grown slightly in the following decades, but foreign travellers remained unimpressed. The British traveller John Carne, who had toured the Levant in the 1820s, explained that Jericho was 'no longer the City of Palm-trees; not a single palm-tree is now to be seen among the few trees that shadow it: its houses are wretched, its situation bad'.[61]

His contemporary Edward Robinson described the village of Ariha as 'the most miserable and filthy that we saw in Palestine', with about 200 permanent residents, and houses constructed of stones taken from ancient ruins. A devout Christian like many of the foreign travellers, he disapproved of the 'illicit intercourse' between the women of the village and those passing through it. Robinson reports that his Bedouin guards despised the inhabitants of the village and sought for 'a paper or written charm, to protect them from the women of Jericho'. For him, the inhabitants of the valley 'retained this character from the earliest ages'. The 'sins of Sodom and Gomorrah', he believed, 'still flourish upon the same accursed soil.'[62] A few years later, John Wilson claimed that the 'present debasement' of the village, and especially

A SITE OF SCIENCE AND EXPLORATION

its women, was due to 'the proximity of a small detachment of Turkish soldiers'.[63]

Perhaps the most negative opinion of all was expressed by Tristram. The women of Ariha came out to put on a show and dance for the party of explorers, but Tristram found them 'miserable and degraded-looking'. Disparagingly, he remarked that the 'women of the Ghor, unlike Moslems of the towns, do not veil, and truly there is no need for them to do so'. His final, objectionable, verdict was that there 'was no trace of mind in the expression of any one of these poor creatures, who scarcely know they have a soul, and have not an idea beyond the day'.[64]

From these scathing accounts we can at least learn that Jericho's economy benefited from sitting astride the trade route going up to Jerusalem, as well as from the presence of an Ottoman garrison. The growing volume of Christian pilgrims meant that, by the turn of the twentieth century, Jericho had modern, three-storey stone houses such as the Jordan Hotel, heralding a period where it held a more positive reputation. Travellers' accounts also reflect how they understood the local class system, with the Bedouins earning more respect than the poor peasantry. However, as the modernisation of the Dead Sea area got underway, the Bedouins themselves were increasingly pushed out of key sites along the lake and forced to abandon their traditional way of life.

EXPLOITATION

During the second half of the nineteenth century, and especially under the reign of Sultan Abdulhamid II (1876–1909), the Ottoman authorities set about modernising the southern Levant. They allowed Circassian and Chechen refugees, uprooted by the expanding Russian Empire, to settle in places such as Amman. Increased settlement led to the revival of the ancient King's Highway, which ran south from Amman through Madaba, across

Wadi Mujib and down towards Aqaba. Wheeled traffic was reintroduced to Transjordan after centuries of absence. In the 1890s, a bridge replaced the ferry across the Jordan River near Jericho, close to where the present-day Allenby Bridge stands. The Ottoman government also began to operate a modest fleet of sailboats on the Dead Sea. Goods and people were transported between the shore nearest to Jericho and Mazraa on the Lisan Peninsula, which offered good access to Karak. The missionary Archibald Forder, who travelled using such a boat in 1895, found that it reduced the length of the trip from Jerusalem to Karak by three days. When the government stopped running the service because of the costs involved, local Arab and Jewish businessmen from Jerusalem and Jericho stepped in, brought more vessels to the lake and increased the traffic of goods and passengers. In preparation for the state visit of the German Kaiser Wilhelm II in 1898, a new road was inaugurated between Jerusalem and Jericho. Jerusalem itself had already been connected by railway to Jaffa on the Mediterranean coast a few years earlier.[65]

Legally, the land around the Dead Sea belonged to Sultan Abdulhamid II. The Ottoman government held the monopoly over extracting and selling crystallised salt from the lake. There were extensive lagoons for this purpose near its northern shore. Soldiers stationed in the area were meant to guard the lagoons and prevent salt being smuggled. However, the physician Masterman noted that, in the rainy season, the soldiers were withdrawn, and the Bedouins plunged into the salt pans and dragged up great masses from the bottom. Obtaining the salt was only half of their clandestine effort: 'The task of smuggling the salt into the towns is one requiring alertness and intrepidity, for, if caught, the Bedouin are liable not only to lose their salt, but also all their camels and mules.'[66]

Around the same time, there was a growing realisation that the Dead Sea area could be developed and exploited much further.

A SITE OF SCIENCE AND EXPLORATION

The first Zionist author to dream of modernsing the Dead Sea, even before Herzl, was the Russian-born Hebrew writer Elhanan Leib Lewinsky. In 1892 he published a utopian description of how Palestine would look in 2040, once it had been developed by Zionist pioneers. Lewinsky's protagonist describes visiting the City of Salt on the shore of the Dead Sea, a bustling urban centre of 100,000 inhabitants. The City of Salt benefits from the booming naval trade across the Dead Sea and up the Jordan River to the Sea of Galilee. It also boasts the High Academy for Earth Sciences, where the wise carry out experiments underground. The city is very beautiful and unique: its roads are paved with asphalt (Lewinsky used the biblical word for Dead Sea asphalt: *hemar*) and the buildings are constructed using blocks of salt rock, thanks to a technological innovation that made it fire- and water-resistant. Because the rock salt resembles crystal, the buildings sparkle – from the sun during the day and at night thanks to electricity.[67] While many of the innovations that Lewinsky envisioned never materialised, his utopia did emphasise one important advantage of the Dead Sea area: it had rare natural resources that could be exploited.

Writing a decade later, Herzl was better briefed about the scientific potential of the Dead Sea. In his utopian novel *Altneuland*, the Zionist leader highlighted that the lake was uniquely rich in potash (potassium-rich salt), a key ingredient in artificial fertiliser. He also pointed out that the Dead Sea salts 'are richer in brome than any other natural lye'.[68] The Zionist movement, focused as it was on a country with very few natural resources, enthusiastically sought to exploit the minerals of the Dead Sea.

Ottoman authorities, too, were not oblivious to the potential of extracting minerals from the Dead Sea. After commissioning a number of expeditions to the area, in 1911 Sultan Mehmed V (1909–1918) issued a *firman* (imperial charter) granting three

Ottoman subjects, Djindjöz Bey, Zuad Bey and Djenab Chehabeddin Bey, the right to extract bromine from the Dead Sea. Hoping to speed up extraction, the *firman* included the condition that the export of bromine had to start within two years. However, despite an extension, by early 1915 the concession-holders had failed to initiate production, leading the Ottoman government to annul the concession.[69]

The person who ultimately succeeded in setting up a large-scale extraction industry in the Dead Sea was Moshe Novomeysky, a young Jewish entrepreneur and mining engineer from Siberia. His initial interest in the Dead Sea was sparked by reading the work of the German geologist Max Blanckenhorn, who had visited Palestine, Syria and Egypt multiple times. In 1911 Novomeysky visited Palestine for the first time. His goal, he later explained, was to study the feasibility of 'extracting the chemical salts in which that lake is almost inexhaustibly rich'.[70] Novomeysky had some experience in extracting chemicals from lakes in Siberia. However, the method he sought to develop for the Dead Sea was entirely new: making use of the region's very high temperatures. He noted that, at different stages of evaporation, various specific minerals should appear in crystalline form. During this first visit, he took some samples of Dead Sea water, experimented with them back in his Siberian laboratory and found that his thesis was correct. But the implementation of the thesis had to wait. It was only after the First World War and the conquest of Palestine by the British that Novomeysky was able to realise his dream and to start a new phase in the history of the Dead Sea.

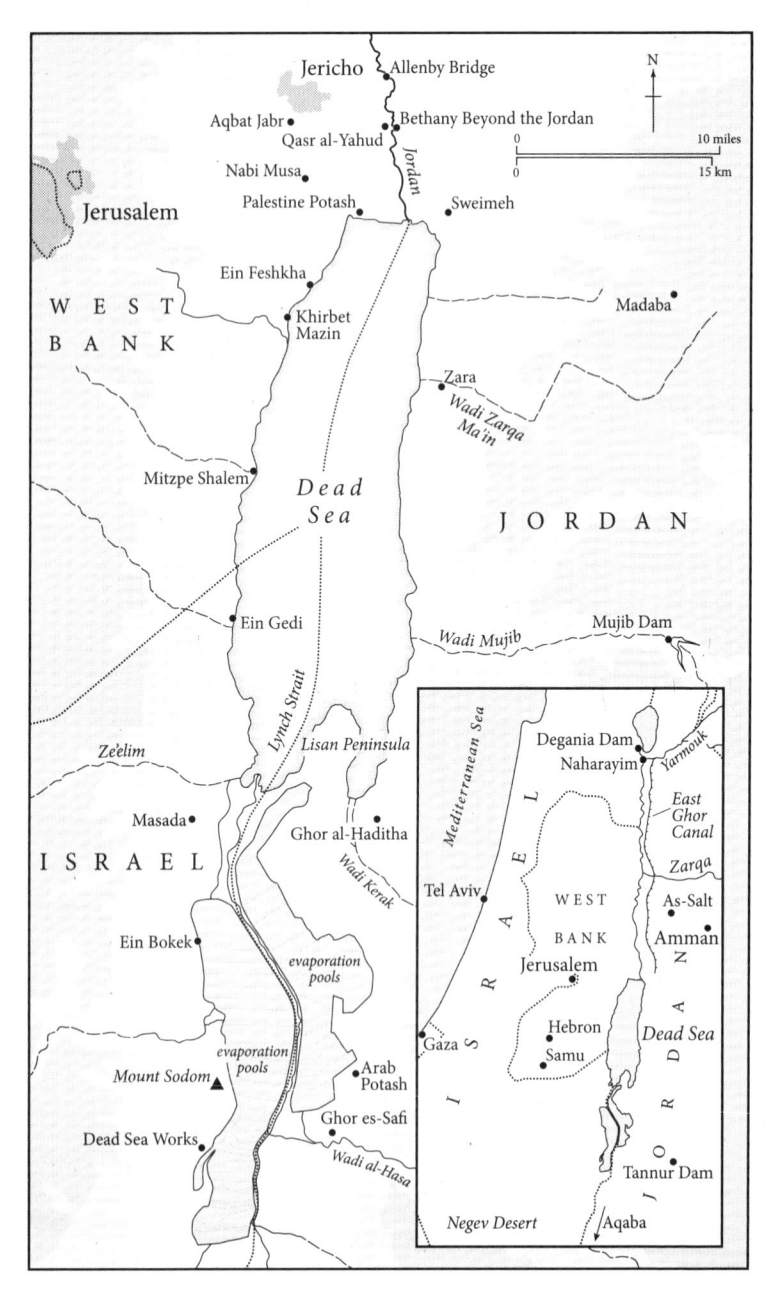

Map 4. The contemporary Dead Sea.

CHAPTER SIX

A SITE OF EXTRACTION, CONFLICT AND WATER SCARCITY

FIGHTING BETWEEN OTTOMAN and British forces near Jericho in autumn 1918, during the final days of the First World War, inadvertently helped bring to light an element of the city's past. The explosion of a Turkish shell exposed the hitherto-buried remains of the Byzantine-era synagogue of Naaran. The veteran French explorer Charles Clermont-Ganneau was able to decipher the synagogue's dedicatory inscription from the exposed portion of the mosaic using a photograph he had received from Major A.M. Furber, a British military officer. Summing up his work a year later, Clermont-Ganneau hailed the Turkish shell, which 'very considerately chose to alight and explode on the very spot where a treasure lay hidden from the eye'.[1]

In the period since the First World War, the Dead Sea continued to be seen in very different ways, depending on one's point of view: a site of religious and historical significance, a site ripe for economic exploitation, a site of potential national renewal. The twentieth-century economic development of the area led to competition over resources between Arabs and Jews. At the same time, the industrial extraction plants that were established by the Dead Sea also provided opportunities for inter-communal cooperation.

Alongside political upheavals and unprecedented industrial activity, this period also witnessed environmental changes as the Dead Sea began to shrink rapidly. The growing populations of the Levant increased the demand on freshwater resources, including those that flowed into the Dead Sea. Before intense human intervention began, the Jordan River and its chief tributary, the Yarmouk, brought between 1,100 and 1,400 million cubic metres (MCM) per year to the lake. The rivers and streams that flowed directly into the Dead Sea from the hills around it also contributed, but a far smaller amount.[2] As we will see, damming and diversions upstream have reduced the flow reaching the lake to a mere trickle. This outcome was the result of a process in which conflict, grand aspirations and economic development all played a part.

FROM OTTOMAN TO BRITISH RULE

On the eve of the First World War, the Ottoman government sought to tap into the resources that the Dead Sea area seemed to offer. In early 1914, Ottoman authorities granted a concession to search for oil near the lake to three local businessmen, Suleiman Nassif, Ismail Hakki al-Husseini and Charles Ayoub Bey, who soon went on to transfer their prospecting licences to the American company Standard Oil. After the Ottoman Empire joined the war in autumn 1914, efforts to further develop the Dead Sea area were intensified so that it could better serve the war effort. Ottoman maps from 1915 show that among the areas of chief interest were those west of Ein Feshkha and south of Jericho. However, wartime conditions meant that actual oil drilling could not take place. Indeed, during the First World War the British Royal Navy placed a blockade on the coastline of the Levant. It seems, however, that Standard Oil, which after the war unsuccessfully pressed British authorities to let them continue exploring for fossil fuels in the area, did not miss out on much. Despite a number of attempts in over a century since the First World War, the Dead Sea area has not yielded substantial oil deposits.[3]

The First World War led the Ottoman authorities and their German allies to upgrade shipping across the Dead Sea so that they would be able to transport grain from the Karak region to Jerusalem, as well as other materiel, more efficiently. They built some barracks and a pier on the northern edge of the lake and introduced a motorboat and a few smaller crafts. The Arabs called this makeshift harbour al-Jadida ('the new'). The Palestinian composer and poet Wasif Jawhariyyeh, who served as a soldier in the Ottoman Army during the First World War, had fond memories from his time drinking, smoking hashish, singing and playing his oud at this base on the shore of the Dead Sea.[4] The British tried to disrupt shipping across the lake through air raids, but with little success. The capture of Jerusalem by General Edmund Allenby in December 1917, a great victory for Britain, made the Ottoman position in the Dead Sea area far more precarious.

In February 1918 a British-led force advanced from the vicinity of Jerusalem eastwards to confront the Turkish forces that still held on to the Jordan Valley, the Dead Sea area and Transjordan. A brigade from New Zealand approached Nabi Musa, where they met initial resistance. However, the Turkish force retreated so as not to be outflanked, while an Australian unit managed to reach the shore of the Dead Sea near Qumran. On 21 February, Jericho, which Turkish forces had abandoned, fell without battle. The British also captured the small Turkish base on the shore of the Dead Sea and took hold of the motorboat, which they later named *Adela* after the General's wife, Adelaide Mabel Allenby.[5] The anchor of the *Adela* is still on display at Kibbutz Kalia, near the lake's north-western shore.

Major operations in Palestine then halted owing to the German offensive of March 1918 which nearly broke through British and French defences on the western front in Europe. General Allenby tried to launch two raids across the Jordan towards Amman, but these were repelled. Between April and mid-September 1918, the main foe that British forces and their allies faced in the area was not Turkish forces but malaria. 'Many of my fellow soldiers fell like the dead and I could not close my eyes', wrote one soldier of the Jewish Battalion stationed near Jericho.[6] Marshland and standing water near the Jordan River and the Dead Sea were a hotbed of malaria-bearing Anopheles mosquitoes. Therefore, General Allenby engaged his troops in a campaign against malaria. An anti-mosquito officer was appointed in every division, and an anti-mosquito squad was set up in every brigade. Thousands of soldiers, along with Egyptian labourers, were employed in drying swamps, diverting streams and marshes into canals, cutting reeds, drying and oiling shell holes and other similar tasks. 'I am campaigning against mosquitoes', Allenby boasted to the Chief of the General Staff in London, 'many acres of bog have been drained and cleared.'[7] However, while in rear areas the situation improved, soldiers stationed on the front line still fell sick because they were exposed to the bites of mosquitoes bred in enemy territory, where no anti-malaria campaign took place.

The offensive against Turkish forces was resumed on 19 September 1918. This advance heralded the end of some 400 years of Ottoman rule. General Allenby's Australian, British, Indian and New Zealand forces moved north along the Jordan Valley and elsewhere in Palestine, achieving victory at the Battle of Megiddo. Meanwhile, Britain's ally from the Arabian Peninsula, the irregular Hashemite force led by the Emir Faisal and assisted by Lawrence of Arabia, advanced along a parallel line in the east, capturing the urban centres of Transjordan. Once Damascus and Aleppo fell in October, the war in the Levant was effectively over.[8] The post-war peace settlement in the Middle East proved complex. Nonetheless, the British victory paved the way for an intense transformation of the Dead Sea area.

EXTRACTION, CONFLICT AND WATER SCARCITY

NOVOMEYSKY'S KINGDOM

The Dead Sea area had long been a source of edible salt from evaporation pools and rock salt from Mount Sodom. However, in addition to sodium chloride (table salt), the Dead Sea is also rich in potassium chloride (potash), magnesium chloride, calcium chloride and bromine, among other compounds. Over the years, these have had several applications. For instance, potash is used in the production of fertilisers. Bromine fulfils a variety of roles, from making engines run more smoothly, to uses in the pharmaceutical industry and in fire retardants in the production of textiles. Magnesium extracted from the lake is nowadays used in the aerospace and automotive industries. Dead Sea minerals are also used in various skincare products. The method of extracting these minerals from the water of the lake was developed over the years and decades since the First World War. Essentially, it relies on precipitation due to water evaporation in a series of evaporation pools. Initially, sodium chloride precipitates and gathers at the bottom of the first pool. Then, the more concentrated brine that remains on top passes into subsequent pools, where the evaporation process continues, until a substance called carnallite is harvested and refined to varying levels to extract the different minerals. But setting up a new extraction industry where none existed before was not a straightforward process.

The British military began to take samples of water from various places around the Dead Sea as early as autumn 1918, before the war was over, to ascertain whether potash extraction would be feasible. Major Reginald Walter Brock, a Canadian geologist, drew up a report which enthusiastically recommended setting up an extraction plant. The economics of imperial rivalry were a key consideration. During the First World War, Germany, with its salt mines at Stassfurt, had 'an unlimited supply of potash salts. Before the war she had a monopoly of the trade'. Major Brock was encouraged not only by the chemical composition of the Dead

Sea's water, but also by the region's climate: 'Evaporation conditions are excellent... the season is long and dry, with clear sunshine, good breeze, strong evaporation with relatively cool nights that facilitate precipitation.' With an eye to alternative sites and potential competitors, he added, 'Conditions are much better than in California, or on the Mediterranean.' He recommended setting up the plant on the northern shore of the Dead Sea, west of the Jordan. According to Brock's report, the Germans had already created a small experimental plant to extract bromine at this site during the war. For this purpose, they had dug a series of what he called 'Hun salt pans'.

Brock's report went beyond technical aspects of extraction to consider other issues such as potential sources of labour. He did not put much faith in the ability of the local population to operate the plant, echoing some of the prejudices of earlier explorers: 'There are a few natives (Bedouins) about Jericho, and east of the Jordan River, and the south east end of the Dead Sea. They are unpromising looking, and have not been trained to steady work, so that not much dependence is to be placed on this source.' Instead, he recommended relying on Chinese labourers, or 'coolies', as he called them. The British Empire had made extensive use of labourers from China during the First World War.

Brock concluded by saying that 'Dead Sea brine is unlimited in amount, and so can form the basis of a permanent industry'. Ensuring a permanent supply of cheap potash, essential for agriculture and other industries, would make the British Empire independent of any foreign source.[9] This last point resonated with officials back in London, who added: 'The importance of Bromine in time of war is well known and it is a fortunate thing that such enormous resources of the element are placed within the control of the British Empire.'[10]

Although the scientific and strategic case for extracting minerals from the Dead Sea was clear, there were a number of political and legal obstacles to overcome. While Britain had conquered the southern Levant in 1917–1918, it wasn't until the early 1920s that the status of Palestine and Transjordan was established. First, at a conference in Cairo in spring 1921, Winston Churchill, then the Minister for the Colonies, sought to reconcile some of the promises Britain had given during the First World War and to solve some of the outstanding political problems of the Middle East. Among other things, he announced that the area east of the Jordan would become the Emirate of Transjordan. This meant it would be separated from Palestine and headed by Abdullah bin Hussein, whose family – the Hashemites – had fought alongside the British during the war. Second, in 1922, Britain formally received the mandate to govern Palestine from the League of Nations.

In June 1922 it became necessary to define the boundary between Palestine and Transjordan so that the respective governments could deal with matters relating to the exploitation of natural resources. The border between the two territories ran in the middle of the Dead Sea. Who, then, should have jurisdiction and oversee concessions in the lake area? It was eventually agreed that the Dead Sea area should be administered under a single mining policy of control and inspection, but that royalties would be divided equally between Palestine and Transjordan.[11]

Next was the thorny question of who should receive the governmental concession to construct and operate the extraction plant. Initially, there were three contenders: the Jewish Russian mining engineer Novomeysky, who had left what became Soviet Siberia and settled in Palestine in 1920; Ibrahim Hazboun, a Catholic businessman from Bethlehem who operated the ferrying service across the Dead Sea; and Major Thomas Gregory Tulloch, who had worked with the Ministry of Munitions during the First World War.

Tulloch, who was probably the first to ask the government in London for a concession, seemed at the beginning to be in the lead. However, Roland V. Vernon, who oversaw the Palestine portfolio at the Colonial Office in London, wanted to see a concrete and detailed plan. He noted that, in Palestine, 'They will remember that many other people from Moses (or at any rate Joshua) onwards have had general ideas about using the Dead Sea, but that nothing very definite has come of it.'[12] With Vernon's encouragement, Novomeysky and Tulloch met and decided to become partners in the bid to extract minerals. Novomeysky initially collaborated also with Hazboun, purchasing some of his boats and undertaking to cover some of the debts the latter had accumulated in his ferrying business. In the mid-1920s, however, Hazboun withdrew from the race because of financial difficulties.

British bureaucracy moved slowly as more contenders entered the fray. In 1927, while deliberations over the technically more complex process of extracting potash continued, the British government granted Shukri Deeb, an Arab businessman from Jerusalem, a concession to mine rock salt at Mount Sodom near the south-western edge of the lake. Salt mining at this site was not new and had been carried out intermittently in previous centuries. Hence, the case for this concession was simpler. Deeb collaborated with, and eventually took over, Hazboun's shipping business. This enabled him to transport everything he quarried from Mount Sodom to the small harbour on the north-western edge of the lake. Deeb's concession included a small plot of land close to the harbour, where some of the rock salt could be crushed. Hence, he was able to produce table salt as well as dark rock salt, which served as part of the diet of horses and cattle.[13]

Finally, in 1929, the Novomeysky–Tulloch partnership secured a seventy-five-year concession for potash extraction, which came into force in January 1930. While Tulloch moved to Palestine, made Jericho his home and played an active role in the Palestine Potash Company till his death in 1938, Novomeysky remained the animating force behind the endeavour. The concessionaries initially received the lease over 4 square kilometres on the north-western edge of the Dead Sea to construct evaporation pools. Soon, lorries bearing building timber, bags of cement, iron piping and other construction material began making their way from Jerusalem down to the Dead Sea. Huts and a new, tiny harbour were built. Evaporation pools were dug and prepared. Water was diverted from the Jordan, both for the workers to consume and for use in the refining process. By February 1931, bromine from the Dead Sea was on the British market.[14]

The granting of the concession did not go down well among nationalist Arab circles in Transjordan and in Palestine. For instance, Shams al-Din Sami of the opposition in the Transjordanian Legislative Council criticised his government for giving its approval for the Dead Sea concession, especially as the project would benefit the Zionists. The nationalist press in Palestine similarly denounced the British decision to award the concession to Novomeysky as a 'robbery'.[15]

However, economic development around the Dead Sea can also tell an alternative story of Arab–Jewish coexistence and cooperation. As extraction around the Dead Sea expanded, Novomeysky assembled a mixed Arab–Jewish workforce. Some workers were drawn from Beit HaArava, a new kibbutz (a socialist-Zionist communal village) that was established near the north-western shore of the lake. Other workers were Bedouin tribesmen from As-Salt and its surroundings in the Jordan Valley and elsewhere in Transjordan. The Emir Abdullah tried to convince the Palestine Potash Company to take on more tribesmen, as employment in Transjordan was scarce. The Company had workers coming not only from neighbouring Jericho and nearby Hebron but also from much more distant places in Syria and the Hejaz. Employment opportunities in the extraction venture led to a growth in the population of Ghor es-Safi. In his memoir, Novomeysky stresses that he always sought 'to see friendly cooperation develop between

Jew and Arab', and that the two peoples 'always worked well together at the Dead Sea'.[16]

One of the first things Novomeysky had to tend to was a local health hazard. As the soldiers who had been stationed near Jericho during the First World War could attest, malaria-bearing mosquitoes were a formidable foe. One known source of malaria was the Jahir spring, about a kilometre north-west of the Dead Sea. The Company dug a direct path for the water to flow to the lake and drained the standing water around the spring. However, even though all nearby breeding places had been eliminated or controlled, mosquitoes continued to infest the residential quarters of the Palestine Potash Company into the 1940s. The chemical compound DDT was introduced in Palestine in the mid-1940s and was used to repeatedly spray springs and freshwater sources such as Ein Feshkha. It was the extensive spraying of DDT that led to the eventual disappearance of malaria cases in Jericho.[17]

In the meantime, the production capacity of the Palestine Potash Company had expanded. In 1933 it was given permission to construct new evaporation pools next to the shore on the eastern (Transjordanian) side of the Jordan River. Furthermore, a new extraction plant was established near the south-western edge of the Dead Sea. The new site on the plain below Mount Sodom offered even higher temperatures and more room for evaporation pools, but these had to be constructed at a safe distance from the lake to allow for fluctuations in its shoreline. A waterpipe was laid down to bring freshwater to the plant from Ghor es-Safi in Transjordan. The southern plant was soon able to outstrip its northern counterpart in potash production. Because there were still no roads running north–south along the shores of the lake, extracted minerals and workers were ferried across the Dead Sea in a journey that took between nine and fourteen hours, depending on the vessel and on weather conditions. In general, the Dead Sea region became the scene of intense economic activity at a scale not seen for centuries. In addition to their other endeavours, Novomeysky, Tulloch, Hazboun and a Dutch dignitary received permission to construct a hotel next to the northern extraction site. They named the resort 'Kalia', after *kalium*, one of the names by which potassium was known.[18]

Having access to Dead Sea potash proved fortuitous for both the government and farmers in Britain following the outbreak of the Second World War and the severing of commercial ties with Germany. 'This war, like the last, reminds us that our supplies of potassic fertilizers come from overseas, mostly Germany and Alsace, now once more in German hands', lamented Sir John Russell, the Director of Rothamsted Experimental Station north of London, in September 1940. However, he noted that 'the position now is certainly better than it was in 1914–1918, since the Dead Sea supplies have been developed'.[19] And indeed, the 1940s saw Palestine Potash Company production scaled up, reaching 100,000 tons of potash in 1941. By 1944, more than half of Britain's potash and 75 per cent of its bromine came from the Dead Sea. Potash from the Dead Sea also dominated the markets in other parts of the British Empire. The market share of the Palestine Potash Company leaped from 10.75 per cent in 1939 to 80 per cent in 1944. Production expanded further after the Second World War came to an end, with 1947 being the best year since the beginning of the concession. However, for the company headed by Novomeysky, it was also the last. Fighting between Jews and Arabs had started in late 1947. In the conflict that followed, the 1948 War, the borders around the Dead Sea would once again change.[20]

CONFRONTATION, PARTITION AND WAR

Much has been written about the history of the Arab–Israeli conflict and about the confrontation between the Zionist movement and the Palestinian national movement in the lead-up to

1948. But how did these tensions and outbreaks of violence affect the Dead Sea area?

An early indication that the Dead Sea region was not exempt from national tensions appeared in summer 1929, as the country experienced its first large-scale wave of Arab–Jewish violence. Moshe (Musia) Langotzky, who was conducting experiments at the Dead Sea for Novomeysky and headed a team of about twenty workers, was tipped off by Arab friends that the workers of Shukri Deeb intended to attack them. Langotzky's group, which included a few Arab sailors, fled on a boat and found refuge in Wadi Zarqa Ma'in on the eastern side of the lake.[21] Despite this episode, from 1930 onwards, Arab and Jewish employees of the Palestine Potash Company worked well together, as we have seen. They resided in two separate but neighbouring camps in the northern extraction site. The company's Board of Directors included Abdul Rahman al-Taji, a Palestinian notable from Wadi Hunayn, south-east of Jaffa. Furthermore, Novomeysky was on friendly terms with Ibrahim Hashem, the Nablus-born politician who served long stints as Minister of Justice and Prime Minister of Transjordan.[22]

But a clash could not be held off indefinitely. In April 1935 the main leader of the Zionist movement and future Prime Minister of Israel, David Ben Gurion, toured the Negev Desert and recognised it as an area of vast potential. He noted that it was very sparsely populated. At the same time, it had extensive ruins, which included the remains of oil presses and wineries, indicating that the Negev had flourished in the past. He also noted that the area around the Dead Sea was 'believed to have rich oil fields ... and other mineral deposits'.[23] The Zionist movement was keen to establish its hold and broaden its reach in southern Palestine and around the Dead Sea. In the mid-1930s key Zionist figures such as future Israeli Foreign Minister Moshe Shertok (Sharett) tried to promote the idea of establishing an agricultural kibbutz south of the Dead Sea, across the Transjordanian border, near Ghor es-Safi. Novomeysky supported this initiative and tried to use his contacts in Transjordan and among British officials to facilitate the sale of land for such a kibbutz, but to no avail.[24]

During the Arab Revolt in Palestine (1936–1939), which was aimed against both the British administration and the Zionist movement, the extraction sites near the Dead Sea remained quiet. However, lorries bearing potash from the Dead Sea were fired upon a number of times on the road leading up to Jerusalem. In January 1939 there was a skirmish between an armed Arab band and members of the Haganah, the main Zionist paramilitary organisation, not far from Mount Sodom. The incident ended with casualties on both sides. Across the country, hostilities died down in the second half of 1939, though not all was quiet near the Dead Sea. In July 1940 the British geologist and former Geological Advisor to the Mandatory Government of Palestine, George Stanfield Blake, was prospecting for oil in a wadi south-west of the Dead Sea when he and one of his police escorts were shot dead.[25]

As we have seen, mineral extraction in the Dead Sea area played a supporting role in the Allied effort during the Second World War. While it never materialised, the region was nonetheless under risk of attack by the Axis powers. In September 1940, the militant exiled Palestinian leader and former Grand Mufti of Jerusalem, Hajj Amin al-Husseini, congratulated the Italian officials he met in Baghdad for air raids conducted by the Italian air force on Tel Aviv (the attack had killed more than 100 civilians). Husseini suggested to the Italians that, among other targets, they should bomb the potash extraction plant on the northern edge of the Dead Sea. He reported that the British were transporting potash through Transjordan to Basra in Iraq, shipping it from there to Britain. The British were likely unaware of the mufti's advice to the Italians, but they recognised the strategic importance of potash extraction and deployed an anti-aircraft battery near the Dead Sea in the early months of 1941.[26]

THE DEAD SEA

Shortly after the Second World War ended, troubles in Palestine resumed, but this time it was the Jewish paramilitary organisations that declared an uprising against the British, aiming to secure the establishment of an independent Jewish state. After trying and failing to come up with a solution that would be acceptable for both the Jews and Arabs of Palestine, the British government decided to refer the problem to the newly created United Nations. In June 1947 a United Nations Special Committee on Palestine (UNSCOP) was dispatched to the country to make its recommendations. Their itinerary included a visit to the Palestine Potash Company site on the northern shore of the Dead Sea, Kibbutz Beit HaArava, the Allenby Bridge across the Jordan and the ancient ruins of Jericho.[27] Members of UNSCOP were divided about what would be the best solution for the country, but most of them recommended partition: dividing Mandate Palestine into two states, one Arab and one Jewish.

The original map submitted by UNSCOP in September 1947 gave almost all of the western shoreline of the Dead Sea to the Arab state. The Jewish state was to receive only the south-western edge, around Mount Sodom. The Zionist and Arab leadership were asked to respond to the plan. The Arabs rejected the map. The Jewish Agency accepted it but asked for certain changes, one of which was the inclusion of the historic site of Masada in the territory allotted to the Jewish state. Negotiations continued for a number of weeks until, in late November, an amended map, which shifted the border near the Dead Sea in accordance with the Zionist request, was put before the UN General Assembly.[28] The Partition Plan was adopted. However, what became known as Resolution 181 almost immediately led to the outbreak of hostilities between Arabs and Jews in Palestine.

The first part of this conflict was largely a civil war, fought between Jewish and Arab citizens of Mandate Palestine. Although the Arab population was almost twice as large as the Jewish population, the latter was better organised. By spring 1948, Jewish forces gained the upper hand, seizing control of a number of key urban areas. Hundreds of thousands of Palestinian Arabs were displaced as a result of the fighting. As the conflict continued, British forces were gradually withdrawing from the country, a process that was set to end in mid-May 1948. The Zionist leader, David Ben Gurion, was pushing to declare the independence of the Jewish state – the name Israel was agreed at the last minute – but knew well that this declaration would be followed by an invasion from the neighbouring Arab states.

As these dramatic events took place, Novomeysky was scrambling to salvage the Palestine Potash Company, especially its northern extraction site that was designated to become part of the Arab state. By April 1948 production had effectively ceased. The withdrawal of British forces meant that the lorries carrying potash from the Dead Sea to Jerusalem could not be escorted. The southern plant too suffered from sabotage. Novomeysky was keenly aware that as soon as the British left, the Arab Legion of Transjordan would sweep in. On 13 May Novomeysky met with the Commander of the Arab Legion, the British General John Bagot Glubb Pasha, as well as other Transjordanian officials on the road from Jericho to the Dead Sea. General Glubb suggested that the concession areas of the company should remain neutral and demilitarised. Novomeysky found this suggestion reasonable but explained that he must consult the authorities in Tel Aviv. He flew to Tel Aviv the next day and met with Ben Gurion. However, the latter was preparing to declare Israel's independence that afternoon, so the two men agreed to meet later that evening. Soon after he parted from Ben Gurion, Novomeysky was involved in an accident. He was hit by a motorcycle and broke his leg. Not only was he not able to return to the Dead Sea area to meet once again with Transjordanian officials, but also circumstances on the battlefield changed. The Arab Legion had crossed the Jordan River and was heading towards Jerusalem. Some of its vanguard units had already captured Jewish settlements in an area called Gosh Etzion. Those left in the northern site of the Potash Company and in neighbouring Kibbutz Beit HaArava found that the Transjordanians were now putting forward far less generous terms. They decided to evacuate en masse and to sail down to the southern site of the company near Mount Sodom under the cover of darkness. Some of the workers sabotaged the extraction site before they left.[29]

Throughout most of the remainder of the 1948 War, the situation around the Dead Sea remained relatively static. Most of the western shores of the lake were under the control of the Arab Legion, while a small Israeli garrison held on to the Potash Company site at the foot of Mount Sodom. Territorial gains by the Arab Legion in what would soon become known as the West Bank meant that the garrison at Sodom was effectively cut off from the rest of Israel for about six months and could only be supplied by aircraft. Likely out of boredom, the commander of the garrison decided on his own accord to launch a raid on Ghor es-Safi, across the border in Transjordan, in June 1948. The attack ended with the retreat of the Israeli force and with casualties on both sides. Soon afterwards the Transjordanians retaliated by shelling the Potash Company site near Sodom.[30] An offensive by the IDF on the southern front in November 1948 managed to break through to Sodom by capturing territory in the Negev and in Wadi Arabah.

In early March 1949, the war was drawing to an end and negotiations on an armistice between Israel and Transjordan were underway. Shmarya Guttman, an archaeologist who during the war served as a major in the IDF, was very keen for Israel to capture the historical sites of Masada and Ein Gedi. These sites were designated for the Jewish state in the UN Partition Plan but were still under Transjordanian control. Guttman made a last-ditch attempt to convince General Yigal Alon, the commander of

EXTRACTION, CONFLICT AND WATER SCARCITY

Israeli forces on the southern front, to set aside a small contingent for this purpose. Alon eventually agreed. There were no roads along the western shore of the Dead Sea. Therefore, an amphibious operation was needed, a very rare occurrence in this conflict. The small IDF force was ferried from the southern end of the lake by a boat that belonged to the Palestine Potash Company. They landed near Ein Gedi halfway through the night between 9 and 10 March 1949. No Transjordanian forces could be seen, but the landing force did encounter a group of thirteen Bedouins. The Bedouins used to grow vegetables in the Ein Gedi oasis where, because of its unique climate, crops ripened earlier than in the rest of the country. Guttman, who had known the group's sheikh previously, convinced the Bedouins to leave the site. The IDF maintained an outpost in Ein Gedi, a few kilometres from the new border, into the 1950s. It was eventually replaced by a civilian settlement, Kibbutz Ein Gedi, which greatly expanded the agricultural capacity of the oasis, taking inspiration from the Jewish village that resided there in antiquity.

The 1948 War had far-reaching consequences for the entire Middle East. The borders of Mandate Palestine had changed. The newly established State of Israel managed to capture more territory than it had been allotted in the 1947 UN Partition Plan. In early December 1948 several Palestinian notables attended the Jericho Conference, which included an audience that was estimated at several thousand. The conference, orchestrated by Transjordanian authorities, expressed a desire for unity between Transjordan and Arab Palestine. Delegates recognised Abdullah as their ruler. They requested that he proclaim himself King and rename his territory, which he soon did.[31] By the time Israel signed the armistice agreement with Transjordan in April 1949, the latter was renamed, becoming the Hashemite Kingdom of Jordan. Its ruler, now King Abdullah I, soon went on to annex the West Bank. Hence, the Palestinian Arab state envisioned by UN

Resolution 181 was not established. Instead, Abdullah offered Jordanian citizenship to the hundreds of thousands of Palestinians who either lived in or came as refugees to the territories under his control.

In terms of the effect of the 1948 War on the demographic composition of the Dead Sea area, the most dramatic change was the influx of Palestinian refugees into Jericho. Ten years earlier, the population of Jericho and the surrounding villages was estimated at just over 4,100.[32] As a result of the war, this number increased more than tenfold. Two refugee camps were established in the area by the International Committee of the Red Cross in 1948–1949. The larger of the two, Aqbat Jabr, was set up south-west of Jericho. For a while, it housed some 45,000 to 50,000 people, who initially resided in tents. Over time, the refugees built mudbrick houses for themselves. The second camp was Ein el-Sultan, north of Jericho, near the ancient Tell. Originally, some 20,000 refugees resided in this camp. In 1950 the running of the camps was taken over by the United Nations Relief and Works Agency (UNRWA) that was set up specifically to cater for Palestinian refugees.[33] The 1948 War was certainly a significant shaper of human history in the Dead Sea area and throughout the southern Levant.

The 1950s and early 1960s saw, on the one hand, attempts to develop the area and to ameliorate living conditions in the Jordan Valley, north of the Dead Sea. On the other hand, the period was also characterised by the persistence of the conflict between Israel and Jordan. An example of attempts at development was the idea to build an airport near Jericho. At the time, the Kingdom of Jordan had two small international airports: one north of Jerusalem and another near the capital, Amman. In 1958 an American aviation mission visited Jordan and recommended building a new airport in Jericho because the ones already operating had limited scope to be extended. An airport in Jericho, it was argued, could help to develop tourism to Jordanian East Jerusalem, 20 miles away, while also being able to take the largest jet transport aircraft of the day. The site near Jericho 'was technically very good with clear approaches and an altitude below sea level although, of course, hot'.[34] The Jordanian government included the Jericho airport in its five-year plan in 1961. However, soon afterwards, the Jordanian official put in charge of airport development decided to enlarge the one in Amman instead. Hence, Jericho's international airport never got off paper, let alone the ground.

At the same time, the conflict between Jordan and Israel continued to simmer, with cross-border infiltrations and raids, including in what had become the wild west of the Dead Sea. For instance, in April 1954, farm animals were stolen from Ein Gedi. In May, an Israeli aircraft spotted the animals on the Jordanian side of the border, in the area near Hebron. This prompted an armed group of ten Israelis to cross into Jordan in an attempt to retrieve the herd. The Israelis encountered local farmers, whom they attempted to question and, later, soldiers of the Jordanian National Guard. In the gunfight that followed, three Jordanian soldiers and one of the Palestinian farmers were killed.[35] In December 1954 two teenage Israelis who were hiking near Ein Gedi went missing. It later transpired they had crossed the border into Jordanian territory, where they were killed. Their remains were discovered several weeks later by an Israeli search party that crossed the border north of Ein Gedi and advanced 7 kilometres into Jordan.[36] In the mid-1960s, the tension between Israel and Jordan manifested itself, also, in the skies over the Dead Sea. In November 1966, during an attack by the IDF on the West Bank village of Samu, Jordanian Hunters and Israeli Mirages entered into a short aerial battle. One of the Jordanian aircrafts was shot down. Its pilot, Lieutenant Muwaffaq Salti, ejected but immediately crashed into the side of a ravine not far from Ein Gedi, while the Hunter flew into the Dead Sea.[37] The period that followed this incident would see not only yet another change in the borders around the Dead Sea but also the beginning of a transformation in the size of the lake itself.

'NO DROP OF WATER SHOULD BE ALLOWED TO RUN TO WASTE'

In December 2022, the Israeli Water Authority published figures showing that rainfall in the four preceding years had been higher than average. Nonetheless, the level of the Dead Sea had dropped by 1.27 metres during that year, a bigger fall than the one registered for 2021.[38] Historically, changes in the level of the Dead Sea were tied to the amount of precipitation in its catchment area, but clearly this was no longer the case. To understand the contemporary decline in the Dead Sea's level, we must turn our gaze some 100 kilometres northward up the Jordan Valley and to decisions made far from the lake itself.

The first large-scale intervention in the flow of the Jordan River came from a concession given by the British Mandatory Government in the 1920s to a former Russian revolutionary called Pinhas Rutenberg. A hydraulic engineer by training, Rutenberg was a member of the Socialist-Revolutionary Party in St Petersburg and took part in the revolutions of 1905 and 1917. However, he was opposed to the Bolsheviks. During the Russian Civil War, he left and eventually settled in Palestine in 1919. Having acquired some experience in dam construction before the First World War, Rutenberg dedicated himself to studying the prospects of, and advocating for, constructing hydroelectric plants in Palestine. His initial plan entailed using all the water flowing into northern Palestine to create two parallel canals – west and east of the Jordan River – to feed fourteen power stations between the Sea of Galilee and the Dead Sea. This ambitious plan had to be modified once the boundaries between Palestine, on the one hand, and Lebanon, Syria and Transjordan, on the other, were agreed in the early 1920s. By 1926, Rutenberg obtained a concession for his Palestine Electric Corporation near the confluence of the Jordan and Yarmouk Rivers. The corporation was to produce electricity for both Palestine and Transjordan.[39]

The construction of dams and canals around the site, which was named Naharayim ('Two Rivers'), began in 1927 and was completed in 1932. The hydroelectric plant made use of the fast-flowing water of the Yarmouk in the winter. Meanwhile, the newly-constructed Degania Dam at the southern end of the Sea of Galilee controlled the outflow to the Lower Jordan River, effectively storing much of the lake's water for several months a year. In the summer months, when the flow of the Yarmouk dwindled, the Degania Dam was opened, with water diverted from the Jordan River feeding the hydroelectric plant through a canal. In this way, electricity was produced at Naharayim all year round.

As the hydroelectric plant began to operate in the early 1930s, a number of kibbutzim (plural of kibbutz) south of the Sea of Galilee started pumping river water to irrigate their fields. Simcha Blass, another entrepreneurial Zionist hydraulic engineer, designed a network of pipes, canals and aqueducts that drew water first from the Jordan River and later also from the Yarmouk, eventually irrigating the whole valley nestled between the two rivers and the Sea of Galilee. The availability of water paved the way for establishing new Jewish settlements in the area and for introducing new agricultural crops.

The use of water upstream had some impact on the Dead Sea downstream, though at this stage it was still relatively minor. The storing of more water in the Sea of Galilee over the winter months for the hydroelectricity plant slightly increased evaporation from that lake. The water pumped from the Jordan and the Yarmouk for irrigation also reduced the amount flowing down to the Dead Sea. The quantities used by the kibbutzim south of the Sea of Galilee, while not inconsequential, were still relatively limited.

When their irrigation network reached full capacity in the 1940s and 1950s, the kibbutzim used between 20 and 40 MCM per year. Meanwhile, the annual flow reaching the Dead Sea was between 1,100 and 1,400 MCM.[40]

Those working by the Dead Sea noticed that its level began to change. The jetty constructed by the Palestine Potash Company near Mount Sodom had to be moved a number of times in the 1930s because of the receding shoreline of the lake, the shallower southern basin being especially susceptible to fluctuations in level. Some change was also visible in the northern basin. Studying the movement patterns of mosquitoes near the western shore of the Dead Sea in 1944, researchers observed that, during the preceding twelve years, the lake had receded. Neev and Emery later calculated that the Dead Sea had dropped by about 3.5 metres during this period. They pointed out that this change was due mainly to human activity, namely damming of the Jordan and Yarmouk and intensive irrigation in the newly established kibbutzim in the Jordan Valley.[41]

However, in the 1940s attention in Zionist circles and among non-Jewish supporters of the Zionist cause was firmly fixed on developing Palestine so that the country could absorb millions of Jewish immigrants. This was happening at a time when the huge scale of the destruction of European Jewry as a result of the Holocaust was beginning to come to light. Walter Clay Lowdermilk, an American soil scientist, proposed a plan to bring Mediterranean water to the Dead Sea to produce electricity, while the freshwater of the Jordan and its tributaries would be diverted to the west and east of the Jordan Valley to extend agriculture. He compared the project to the construction of the Boulder Dam in the United States and suggested forming a Jordan Valley Authority, similar to the Tennessee Valley Authority, a public-works agency that had successfully harnessed rivers for irrigation and hydroelectricity.[42]

Lowdermilk's plan was enthusiastically picked up by the Zionist leader and future President of Israel Dr Chaim Weizmann, who recruited more experts in the US and in Palestine to develop it further.[43] A memorandum drawn up by these experts in summer 1944, titled 'Plans for irrigation and hydro-electric development in Palestine', sets out the scope of their ambition. The plan's 'guiding principle is that no drop of water should be allowed to run to waste'. Therefore, a network of dams and canals should harness Palestine's rivers, as well as make efficient use of runoff, leading to 'a virtual transformation of the country by endowing it with numerous "artificial streams", watering arid and semi-arid regions'. Echoing ideas expressed in Herzl's *Altneuland*, the plan also considered making use of the 'unusual topography' of the country, namely the great difference in levels between the surface of the Mediterranean and the Dead Sea: 'Since the irrigation program calls for diversion of Jordan River water for irrigation, it is proposed to replenish the Dead Sea by diverting into it salt water from the Mediterranean [in] quantities equivalent to the loss caused by the diversion of Jordan waters.' Hence, the authors suggested constructing a 95-mile system of canals and tunnels starting near Haifa on the Mediterranean coast and terminating in the Dead Sea. En route, the water would be used to produce energy. The availability of cheap energy would, in turn, assist in the exploitation of the Dead Sea, 'which contains enormous quantities of minerals in solution', including some '22 billion tons of magnesium salts'. Developing Palestine's agriculture through irrigation and substantially increasing its capacity to produce energy, the authors argued, would enable its farming population to double and, possibly, treble the overall number of people residing in the country. This thrust of development would benefit not only the Jews, but also the region's Arab inhabitants:

> One of the objectives of the project is to raise the extremely low standard of living of the Arab peasantry and reduce the

present gap between their primitive agriculture and the advanced economy of the Jewish section of the population. It would also substantially benefit Trans-Jordan.[44]

As is often the case with ambitious plans, only a small part of the ideas put forward in the memorandum were realised.

Shortly after Israel was established, the hydraulic engineer Blass was charged with leading the planning of a National Water Carrier – a system of giant pipes, open canals, reservoirs and large pumping stations – that would bring water from the Jordan River basin to the arid western Negev. Developing the Negev was a priority for Prime Minister Ben Gurion. A few years earlier, Blass had come up with his own plan to bring water to the Negev: constructing a canal and tunnel that would run more or less parallel to the Jordan Valley before turning south-westwards towards the Negev. However, by the end of the 1940s, his ideas needed refining in light of the changing geopolitical situation. Given Jordan's control of the West Bank following the 1948 War, a new route had to be devised. Planning the National Water Carrier proved incredibly complicated, not only from an engineering point of view, but also strategically. Initially, in the early 1950s, Israel hoped to draw water at a high elevation from the Upper Jordan River, before it flows down into the Sea of Galilee. However, the area chosen for a dam on the Upper Jordan River was on the border between Israel and Syria. Still disputed, it was often the site of skirmishes. Israeli attempts to dam the Upper Jordan River in 1953 led to a diplomatic row in which the US had to intervene. Subsequently, the Israelis decided to draw water from the Sea of Galilee instead.[45]

Israel's National Water Carrier remained contested diplomatically and militarily – the first ever armed attack by the Palestinian organisation Fatah was an unsuccessful attempt to sabotage it. Nonetheless, the construction of the Carrier, then Israel's largest infrastructure project, was complete by late 1964, and it began to operate in early 1965. To provide sufficient water for the Carrier, the flow of water through the Degania Dam was reduced, thereby artificially raising the level of the Sea of Galilee. Within a few years the flow into the Lower Jordan River dropped from around 480 MCM a year to a mere 40 MCM. This had a notable impact downstream. Before the National Water Carrier was inaugurated, the level of the Dead Sea was -398.7 metres.[46] The lake's level has been dropping ever since, partly because of Israel's actions and partly because of the actions of neighbouring Jordan and, to a lesser extent, Syria.

POPULATION GROWTH, DISPLACEMENT AND SOARING DEMAND FOR WATER

When it became independent in 1946, Transjordan – as it was known at the time – had a small population of approximately 350,000 people. Successive waves of refugees from Palestine (1948), the West Bank (1967), Iraq (1990 and again in 2003) and Syria (post-2011), alongside high birth rates, have increased the population of Jordan in leaps and bounds. In 2024 it is estimated at more than 11 million. As Jordan is mostly an arid country, with 90 per cent of its area receiving average annual rainfall of less than 200mm, the growing population has placed an immense pressure on its freshwater resources. Jordan also has an agricultural sector that depends on irrigation to increase its crop yields and vary the number of crops that can be planted each year. While rain-fed land usually produces only winter crops, irrigation paves the way for the introduction of high-income crops. The need to sate demand for water has, over time, contributed to the dropping level of the Dead Sea.

A plan focusing specifically on irrigating the eastern bank of the Jordan Valley by diverting water from the Yarmouk River first emerged in 1938. Michael G. Ionides, the Director of Development for the Government of Transjordan, proposed the construction of such an irrigation canal, running south along the east bank of the Jordan, thereby enabling the transformation of between 50,000 and 100,000 acres into agricultural land.[47] Not much was done with this proposal until the 1950s, when the dispute over Jordan River water between Israel and Syria, as well as water-related tensions between Jordan and Israel, led to the Eric Johnston Mission of 1953–1955. Johnston, a Special Representative appointed by US President Dwight Eisenhower, sought to gain Israeli, Jordanian, Syrian and Lebanese agreement to a plan for the unified development of water in the large catchment area of the Jordan River. He managed to secure an informal agreement on some technical aspects of the Jordan Valley Plan he devised, including the proportion of Yarmouk and Jordan River water that Israel and the Kingdom of Jordan would be entitled to. No formal agreement was signed, but the allocation of water that Johnston devised served as a basis for negotiations in the decades that followed. Recognising Jordan's irrigation needs, Johnston envisaged the Hashemite Kingdom as the principal beneficiary of the Yarmouk River, with Syria, where the 47-kilometre-long river originates, and Israel downstream each being entitled to smaller allocations of water.[48]

Middle Eastern tensions in the late 1950s prompted the United States to support the construction of what became known as the East Ghor Canal. King Hussein of Jordan, the grandson of Abdullah I, was an ally of the West in the Cold War. However, Hussein was increasingly coming under pressure due to the soaring popularity and influence of his regional rival, President Gamal Abdel Nasser of Egypt. In early 1958 Syria and Nasser's Egypt declared that they would join together to create one state, the United Arab Republic. To buttress King Hussein's regime, US Secretary of State John Foster Dulles offered the Jordanians economic and technical assistance in constructing the first stage of the East Ghor Canal. King Hussein envisaged the canal eventually bringing 150,000 acres of land under intensive cultivation, and even producing electric power, though the latter ambition did not materialise. The Jordanians reassured the Americans, and through them the Israelis, that the quantity flowing through the canal would not exceed their Yarmouk water allocation based on Johnston's Plan. An Italian firm, Società Imprese Venete per Costruzioni all'Estero, was hired in 1959 to construct the northern phases of the canal. By October 1961, King Hussein officially opened the floodgates of the first part of the East Ghor Canal, a 23-kilometre stretch out of 70 that were planned. It was the largest development project that Jordan had undertaken since it gained its independence.[49]

At the time, about one third of Jordan's population was composed of Palestinian refugees that the kingdom had absorbed following the 1948 War. In August 1949 Prince Abdul Majid Haidar, Jordan's Minister in London, published a letter in *The Times*, calling for the Jordan Valley to be developed urgently 'because of the vast number of Arabs, both Muslim and Christian, who have been forced by events to leave their homes and seek refuge in the Kingdom of Jordan'. He added that 'it is only by irrigation that the potentially rich land bordering the river Jordan can be made suitable for settlement on a large scale'.[50]

The United States was also keen for Jordan to resettle some of the Palestinian refugees in the Jordan Valley. Eric Johnston promoted this idea, noting in June 1957 that irrigating and developing the valley would offer 'the opportunity of permanent livelihood to some 175,000 persons, who, in large majority, would be Arab refugees'.[51] By the time the first part of the East Ghor Canal was opened in 1961, the Jordanian government had settled some 18,000 Palestinian refugees in the Jordan Valley. The first steps were not without difficulty. Initially, the vast majority of farmers who participated in the project to develop the Jordan Valley lived in mudbrick houses and had to obtain their drinking water from the Canal itself. Nonetheless, alongside irrigation, more modern farming methods were introduced, and soon the land's levels of productivity rose, with the yields for wheat, tomatoes, cucumber, squash and watermelons doubling.[52] Some of these crops require significant amounts of water and could not be grown in the Jordan Valley without the help of irrigation.

The Six-Day War in June 1967 and its aftermath severely hindered the agricultural development of the eastern side of the Jordan Valley. In this short conflict, Israel fought against Egypt, Syria and Jordan. Among the vast territories that Israel conquered was the entire West Bank. The border between Jordan and the territory controlled by Israel moved to the Jordan River and cut through the Dead Sea, as it had done during the British Mandate. The Israeli conquest triggered a new wave of Palestinian refugees from the West Bank. For instance, as a result of the Six-Day War, the number of Palestinians residing in the West Bank refugee camp of Aqbat Jabr near Jericho dropped from several tens of thousands to around 2,700, with many fleeing eastwards, across the Jordan, to the Hashemite Kingdom.[53]

Between 1968 and 1970 the Jordan Valley border zone became the scene of raids on Israel by Palestinian militants, who used the Kingdom of Jordan as their main base of operations, and counter-attacks by the IDF into Jordanian territory, east of the Jordan River. One place that suffered from this period of low-intensity conflict was Qasr al-Yahud, the traditional baptism site on the Jordan, north of the Dead Sea. In the 1920s and 1930s the Catholic Church and other Christian denominations had bought plots of land on the western bank of the river. Subsequently, seven churches and monasteries were built in the vicinity. These establishments were abruptly abandoned following the 1967 Six-Day War and the Israeli conquest of the West Bank. Pilgrimage to the site all but ceased. A few Ethiopian monks still hung onto their monastery, which had been established by Empress Menen Asfaw, wife of Haile Selassie. However, a ceremony held by the monks came under fire from the eastern bank of the river in spring 1969.[54] Shortly afterwards, the Israeli military laid landmines around the churches and monasteries to prevent infiltrations by Palestinian organisations from the eastern side of the Jordan. Consequently, the Christian sites of worship on the western side of the river remained inaccessible for several decades.

Further north, many of the farmers who had settled in the Jordan Valley to benefit from the East Ghor Canal had to flee their homes in 1968. Between 1968 and 1970, the IDF shelled the canal several times, prevented the irrigation of thousands of acres of Jordanian farmland and demolished many buildings in an attempt to eliminate *fedayeen* (Arab guerrilla) safe havens. One operation in June 1969 saw an Israeli commando unit cross the border into Jordan and detonate explosives that breached the East Ghor Canal near the tunnel that diverted water into it from the Yarmouk River. After some repair works, the canal was struck again in August 1969, this time by the Israeli Air Force. Consequently, the canal was out of operation, with water flowing back into the Yarmouk. In response, Palestinian militants bombed the turbine building of the Naharayim hydroelectric plant. Although the plant was within Jordanian territory, had been damaged and stopped working during the 1948 War, it was still the site of a small reservoir. Due to the explosion, water was released and flowed down the Jordan, making it more difficult for a nearby Israeli kibbutz to pump water for its fields.[55] Ironically, the only beneficiary from the violence in the Jordan Valley was the Dead Sea, which for a while received more water than it would have had the Jordanians and Israelis been able to irrigate their farmlands without disturbance.

Cross-border raids by the Israelis and the growing power of the Palestine Liberation Organization (PLO) within Jordan were beginning to challenge the authority of King Hussein's regime. In autumn 1970, Hussein began a large-scale effort to suppress the Palestinian organisations that operated within the kingdom. Following these clashes, which the Palestinians remember as Black September, the Palestinian leadership left Jordan, moving its headquarters to Lebanon. While raids by Palestinian militants across the Jordan River into Israel decreased, tensions between Israel and Jordan over diverting water from the Yarmouk persisted until the late 1970s.[56]

Tensions subsided from July 1979 onwards, as secret meetings began between Israeli and Jordanian officials to mitigate conflict by dividing and monitoring the flow of the Yarmouk in a manner that both sides could accept. In September a weir was constructed across the Yarmouk, topped with gravel-filled bags that could be adjusted to divert the river's flow between the two riparian states. Subsequently, a few meetings took place each year, and the issue most often discussed was summer water allocations. As there were no formal diplomatic relations between the two countries, Jordanian and Israeli water experts met under the auspices of the UN either at a Jordanian picnic table on the Yarmouk's bank near the East Ghor Canal intake or at an Israeli shed by the Allenby Bridge. Once a modus vivendi was established, Jordan was able to divert approximately 130 MCM of Yarmouk water into the Canal each year, while Israel drew a further 45 MCM for the fields north of Naharayim. As a result of these diversions, the historical flow of the Yarmouk into the Jordan River and onwards to the Dead Sea was substantially reduced.[57]

Unfortunately for the level of the Dead Sea, demand for water in this arid part of the world continued to outpace supply. Hence, the following decades saw the construction of more dams and new diversions in the catchment area of the lake. In 1977–1978 a Yugoslav company constructed the King Talal Dam on the Zarqa River, which flows westward, passing between Jerash and Amman, into the central Jordan Valley. The dam created one of the largest reservoirs in Jordan, which catches both floodwater and treated (or barely treated) wastewater. The water in the reservoir is used to generate electricity and to irrigate some 15,000 acres of farmland.[58]

As the Yarmouk rises in south-western Syria, itself an arid country, the Syrian government also sought to utilise its water. Since the mid-1980s more than forty dams were built in the upper Yarmouk basin in Syria, removing some 200 MCM by the 2000s. In 2003–2004 a joint Jordanian–Syrian project that had been discussed intermittently for decades began to capture even more water in the Yarmouk basin: the al-Wehda Dam ('Unity Dam', formerly known as the Maqarin Dam). The water gathered here creates a reservoir, through which the international border passes. Hence, the water is shared between the two countries, though tensions remain. For instance, in 2012, the Jordanian Water Minister, Mousa Jamani, accused the Syrians of violating previous water-sharing agreements by drilling 3,500 wells in the Yarmouk's basin.[59]

Finally, even most of the rivers and streams that in the past emptied directly into the Dead Sea, contributing between 250 and 350 MCM per year, are today captured and used for consumption or irrigation.[60] The Mujib Dam was constructed in the early 2000s to help relieve water scarcity in Amman. Some water is released through the dam into the Wadi Mujib nature reserve but is siphoned and diverted by a pipe on its western end. Only when there are flash floods in the winter does a substantial amount of water reach the Dead Sea through this beautiful canyon. The same period saw the construction of the Tannur Dam and the creation of a reservoir in Wadi al-Hasa further south. In 2022 yet another dam east of the lake was completed. The Ibn Hammad Dam near the village of Ghor al-Haditha creates a small reservoir that diverts water for irrigation and industrial uses. Across the

Dead Sea on the Israeli side, the rivers and springs in the Ein Gedi oasis are also captured and used for various needs, including to supply Kibbutz Ein Gedi and irrigate its fields. A pipe leads water from one of the streams directly down to the Ein Gedi Mineral Water plant, where it is bottled and sold across Israel.

Historically, the level of the Dead Sea was sustained, with some fluctuations, by the combined flow of the Jordan, Yarmouk and Zarqa rivers, as well as the rivers and springs around the lake. Since the 1960s, much of the water that had previously flowed down the Jordan River was diverted to the northern Negev by the Israeli National Water Carrier. Meanwhile, settlements along the eastern bank of the Jordan Valley were supplied by the East Ghor Canal. In the 1980s, it was extended to the south, reaching a length of 110 kilometres, and was renamed the King Abdullah Canal. Today, the Dead Sea receives less than 10 per cent of its historical inflow. Instead, water is consumed by a population that has risen severalfold in the past eighty years, as well as being used to irrigate agricultural lands that would otherwise be virtually unusable.

THE COMPENSATION THAT NEVER CAME

The planned diversion and hydroelectric schemes devised in the first half of the twentieth century by Herzl, Rutenberg, Lowdermilk and Blass all sought to 'compensate' the Dead Sea: while the freshwater of the Jordan and Yarmouk Rivers would be diverted, Mediterranean seawater would be brought to the Dead Sea through canals to produce electricity, enabling the lake to retain its level. However, the plans to compensate the Dead Sea remained dormant for a number of decades, even as Israel's National Water Carrier, which took water away from the basin, was constructed and began to operate. Then, another Arab–Israeli war brought the dormant idea back to light.

EXTRACTION, CONFLICT AND WATER SCARCITY

The war between Israel, on the one hand, and Egypt and Syria, on the other, in October 1973 saw not only intense fighting in the Sinai Desert and Golan Heights, but also an oil embargo imposed by several Arab states on Israel's Western allies. This embargo soon led to an international energy crisis. As a country with very few natural resources, Israel relied heavily on imported fossil fuels to meet most of its energy needs. In the words of Yuval Ne'eman, the physicist-turned-politician who became deeply involved in addressing the problem, the 'seawater-for-energy half' of Herzl's plan 'was left undone, as it could not compete with the low oil prices then prevalent', but this changed in 1974.[61] The Israeli government commissioned a study of the feasibility of what came to be known as the 'Inter-seas Water Conduit'. Ne'eman chaired the project's Steering Committee, which submitted its recommendations to the government, and the latter accepted these in principle in August 1980. The hydroelectric project sought, in the first instance, to refill the Dead Sea to its 1950 level of -393 metres. After that, the water supply was expected to be in equilibrium with evaporation. The Steering Committee was aware that introducing large amounts of seawater into the Dead Sea would dilute the lake's potash content and could decrease potash production at the extraction sites in the south. However, they pointed out that there were 'ecological, touristic, historical, and other values of the preservation of the Dead Sea at its 1950 level'.[62]

The Steering Committee considered seven possible routes: six connecting the Mediterranean and the Dead Sea, and one connecting the Red Sea and the Dead Sea. To avoid disturbing underground aquifers, and for economic reasons, the Committee indicated that the route from the Gaza Strip to Ein Bokek on the western side of the Dead Sea's southern basin 'presents the best answer to the environmental-ecological-hydrological-geological-tectonic issues' it had considered.[63] They argued that 'political complications' could be minimised by passing a 7-kilometre-long buried conduit underneath the Gaza Strip. The Israeli plan in general, and the route going through the Gaza Strip in particular, were harshly criticised by the neighbouring Arab states.

The Arab League condemned Israel's intention to build a canal linking the Mediterranean Sea to the Dead Sea, arguing the project would 'bring about fundamental and irreversible geographical, demographical, ecological, environmental and economic transformations, creating great damage to the vital interests and rights of the whole region, and in particular, those of the Hashemite Kingdom of Jordan and the inalienable rights of the Palestinian people'.[64] The Israeli diplomat Yehuda Zvi Blum tried to ward off criticism from Jordan in particular. He argued that Israel only sought to restore the Dead Sea to its level in the 1950s and highlighted the 'great potential this project holds in store for both countries':

> Since this project can also be highly beneficial to Jordan, it is both astonishing and disappointing that instead of welcoming such an endeavour, the Permanent Representative of Jordan has found it necessary to call it a 'fiendish plan'. In view of the use by the said Representative of derogatory and pejorative terms concerning the Israel project, it is also somewhat surprising that the Government of Jordan should be planning the construction for similar purposes of a canal of its own linking the Red Sea and the Dead Sea.[65]

Despite the arguments put forward by Israel, the Arab League states enjoyed widespread support in the UN General Assembly, which passed a resolution in December 1982:

> Recognizing that the proposed canal, to be constructed partly through the Gaza Strip, a Palestinian territory occupied in 1967, would violate the principles of international law and affect the interests of the Palestinian people ... the canal

linking the Mediterranean Sea with the Dead Sea, if constructed by Israel, will cause direct, serious and irreparable damage to Jordan's rights and legitimate vital interests in the economic, agricultural, demographic and ecological fields.[66]

Eventually, it was not criticism from the Arab League or condemnation from the UN that thwarted the 'Inter-seas Water Conduit' project. The 1980s saw a sharp decline in oil prices. These made the vast investments needed to construct the canal far less appealing. Consequently, the Israeli plan was shelved.[67] It wasn't until the 2000s that proposals to bring seawater to the Dead Sea resurfaced. By then, the environmental crisis in the Dead Sea was plain to see.

CHAPTER SEVEN

※

A SITE OF RUIN, BEAUTY, HEALTH AND HOPE

IN JANUARY 2024, the level of the Dead Sea fell to below -438.5 metres. Currently dropping at a rate of approximately a metre per year, at the time of writing the lake is about 40 metres lower than it had been in the mid-1970s. Before large-scale diversions began, the Dead Sea covered an area of around 950 square kilometres. By 2021, it had shrunk to between 600 and 650 square kilometres.[1] In large part, the shrinking of the Dead Sea is the result of the drastically reduced amount of water that flows into it. As fellow Dead Sea historian Orit Engelberg-Baram points out, there was nothing unexpected in the lake receding. Indeed, in the 1960s the reduced volume of the Dead Sea was not even thought of as a problem.[2]

Another contributor to the woes of the Dead Sea today are the large mineral extraction sites that take up the entire southern basin of the lake on both its Israeli and Jordanian sides. This industry not only dramatically expanded in size since the days of Novomeysky but also artificially enlarged the surface area of the lake, thereby increasing the rate at which its waters evaporate. In fact, were it not for a system of replenishment to top up the level of the evaporation pools in the area, the shallower southern basin of the Dead Sea would likely be completely dry today.

THE DEAD SEA

The Dead Sea is not the only lake to suffer from a severely diminished inflow of water. Since the mid-twentieth century, this has been the fate of several other lakes around the world, including the Aral Sea in the former Soviet Union, Lake Urmia in Iran, Lake Chad in west-central Africa and the Great Salt Lake in Utah, USA. Shrinking lakes can create health hazards. For instance, the large expanse of dry lakebed sediments left behind by the receding Aral Sea has led to dust storms that negatively affect people's health in surrounding areas. As we shall see, the retreating Dead Sea created a different hazard that communities and companies operating around the lake have to contend with.

Alongside mineral extraction, the Dead Sea sustains a sizeable tourism industry and a market for health products and treatments. Yet while the economy of the Dead Sea area gained significantly from the peace process of the 1990s, both the local economy and peace in the region are under threat at present. If the dire state of the lake, and the uncertainty of businesses that depend on it, is to be ameliorated, the peoples of the region will need to work together.

EXTRACTION AT ALL COSTS

In 1952 the Palestine Potash extraction site near Mount Sodom was nationalised by the Israeli government. The new company was given the rather uninspiring name 'Dead Sea Works'. Novomeysky lost control of his former kingdom and formally retired in 1953. From 1955 onwards the Dead Sea Works was led by General Mordechai Maklef, who had recently finished serving as the Chief of Staff of the IDF. Maklef was well aware of the plans to divert water away from the Jordan River through the National Water Carrier. He was also aware that this diversion would lower the level of the Dead Sea. But despite the expected loss of some of the water needed to sustain extraction, Maklef saw an economic opportunity in the level of the lake dropping and its shallower southern basin drying up. The Dead Sea Works sought to enlarge their evaporation pools so that they could extract more potash. 'Treat the sea as one huge quarrying site where work has barely begun', Maklef told one of his engineers. He envisaged turning the whole Israeli side of the southern basin into a series of large evaporation areas, separated by dikes and connected through canals and pumping stations.[3]

The major restructuring of the environment that this plan entailed required significant funding, something the coffers of the Israeli government were ill-equipped to provide. Therefore, the government approached the World Bank and asked for a loan. The scale of the endeavour was succinctly summarised in the report and recommendations drawn up by the World Bank in 1961: 'This extension will increase capacity by about three times to a total of about 600,000 tons [of potash] per year. Approximately 60 kilometers of dikes will be constructed and about 100 square kilometres of the Dead Sea enclosed for use as concentrating pans.' The Bank assessed that 'the Project is soundly conceived technically and the market prospects are favorable', noting that 'Potash, one of the three basic fertilizers, is expected to be increasingly required, particularly in the areas of Asia and Africa.' As a result, it granted the project a loan of $25 million, a considerable sum of money at the time.[4]

To pave the way for the World Bank loan, the Israeli government gave the Dead Sea Works a generous concession that covered an area of 650 square kilometres around the south-western edge of the lake. The transformation of the Israeli side of the southern basin began in the mid-1960s. Huge quantities of earth were quarried from the concession area around the lake to build the dikes. By the late 1960s, a long dike was constructed from the southernmost edge of the lake northward, enveloping the Israeli section of the southern basin and marking the international border with

Jordan. Gradually, the western part of the basin was divided into a series of large pools. Then, in 1977, the level of the lake dropped below the height of the Lynch Strait that separates the Dead Sea's northern and southern basins. To ensure its evaporation pools could be replenished, the Dead Sea Works constructed a pumping station at the southern edge of the northern basin of the lake, east of Masada. To transport the water across the dried-up Lynch Strait to the evaporation pools in the southern basin, the company constructed a 12.5-kilometre canal. Once every few years since, the falling level of the northern basin has forced the company to move the pumping station further north, nearer to the retreating shoreline.[5]

In the 1990s the concession given to the Dead Sea Works was amended, while the company passed into private hands. Today, the extraction site near Mount Sodom is run by Israel Chemicals Ltd (ICL), one of the world's largest producers of potash and bromine. To fill its evaporation pools in the southern basin, ICL pumps more than 430 MCM per year from the lake's northern basin. In 2024, the Israeli Water Authority estimated that the water taken by ICL from the Dead Sea's northern basin lowers its level by approximately 28cm annually.[6]

A Jordanian company seeking to extract potash from the eastern side of the Dead Sea was formed in 1956. However, it wasn't until 1977 that the Jordanian government decided to develop the area between the Dead Sea and Aqaba. In addition to constructing a new highway running down to the Red Sea coast, the long-discussed plan to establish an extraction plant near the south-eastern shore of the Dead Sea was finally implemented. Like the Dead Sea Works on the Israeli side of the lake, the Arab Potash Company established its facilities in the southern basin and around its shore. Once preparations for construction began, archaeologists quickly sought to survey the site designated as the 'township' for the workers of the Arab Potash Company because

A SITE OF RUIN, BEAUTY, HEALTH AND HOPE

it was right next to the Early Bronze Age cemetery of Bab edh-Dhra (they found remains and artefacts dating from the Neolithic to the Byzantine era). Potash production began in 1983 and followed similar methods to those used by the Dead Sea Works on the opposite side of the lake. The Arab Potash Company built a series of dikes with a total length of 117 kilometres. These divided the Jordanian side of the southern basin of the Dead Sea – an area of 150 square kilometres – into evaporation pools.[7] To transport water from the northern basin to the extraction site, the first pool, along with a connecting canal, had to be constructed on the western foot of the Lisan Peninsula, on part of what once was the Lynch Strait. Today, the Arab Potash Company prides itself on being the eighth-largest potash producer in the world.

As contemporary satellite images show, the entire southern basin of the Dead Sea has been transformed into a massive extraction site. There are some thirty large and quite a few smaller evaporation pools on the Israeli side of the border. These extend over an area that stretches for 30 kilometres, from the Lynch Strait in the north to a point about 13 kilometres south of the historical edge of the Dead Sea. The slightly smaller extraction site on the Jordanian side of the border consists of some twenty large evaporation pools, nestled between the Lisan Peninsula in the north and Ghor es-Safi in the south. At present, it is estimated that the water taken by the Arab Potash Company from the northern basin of the Dead Sea lowers its level by about 22cm per year. However, in March 2024 Israeli parliamentarians were told of a plan to expand the evaporation pools on the Jordanian side of the lake, giving them an extra capacity of 60 MCM, and potentially increasing the amount of water the Arab Potash Company draws.[8] The extraction industry, on the one hand, and the diminished amount of water reaching the Dead Sea, on the other, have completely altered the lake's environment.

THE DEAD SEA

THE DYING DEAD SEA

The drop in the level of the Dead Sea has had a number of consequences. One of these has been a change in the composition of the lake's water. When Neev and Emery conducted their research in 1959–1960, the Dead Sea was still a meromictic lake with two distinct layers of water. The top layer, or upper water mass, extended from the surface to a depth of about 100 metres. The layer was warmer, lighter and less salty. Despite the intense evaporation, it was sustained by the inflow of water from the Jordan and other springs and rivers around the lake. The bottom water mass was colder, denser, heavier and more saline. Change came in the second half of the 1970s, which saw decreased rainfall in addition to a reduced inflow of water caused by damming and diversions upstream. The two layers began to mix and, by 1979, they had fused into one.[9]

The dropping level of the lake has also had an impact on the microorganisms that Volcani had identified in the 1930s and 1940s, when the overall salinity was much lower and the upper water layer had less than 80 per cent of its present-day salt concentration. The last time a large-scale appearance of *Dunaliella* algae was recorded was in the winter of 1992, which saw unusually high levels of precipitation throughout the region. Heavy rainfall in the hills around the lake and floodwater from the Yarmouk, alongside a partial opening of the Degania Dam on the Jordan, brought about a short-lived rise of 2 metres in the level of the Dead Sea, an extremely rare occurrence. This huge influx temporarily recreated an upper, less saline, layer in the lake's water. 'At the beginning of May,' observed geologist Eli Raz of Kibbutz Ein Gedi, 'the Dead Sea acquired a green color, and by the end of the month [the water] turned reddish.'[10] Later simulations illustrated that a dilution of the upper water layer with more than 10 per cent of fresh water is required to initiate a mass development of algae.

However, once the salinity becomes too high even for the *Dunaliella* cells, they rapidly disappear from the water column.[11] The rising salinity of the Dead Sea makes it well and truly dead.

If we lift our gaze up from the water, the retreating shoreline of the Dead Sea has also affected the area's human population. While the expected drop in the level of the lake was perfectly understood in the 1950s and 1960s, one of its most dangerous side effects came as a surprise: the appearance of sinkholes. Underground clusters of salt sediments, which had been left behind during the Pleistocene by the predecessors of the Dead Sea, remained in place so long as the Dead Sea's level was more or less stable. However, once the shoreline retreated, the underground clusters began to dissolve as fresh water from the surrounding hills flowed deeper underground before reaching the Dead Sea. As a result, subsurface cavities formed, and the earth above the cavities began to collapse.

The scale and severity of the problem was not immediately understood. Flash floods occasionally caused sinkholes to form even before the level of the lake began to drop. One had damaged the road between Mount Sodom and Ein Gedi after heavy rains in April 1964.[12] But since the early 1980s, sinkholes began to appear more frequently, sometimes in clusters. For instance, contractors building dikes for the evaporation pools of the Arab Potash Company near the Lisan Peninsula found that sinkholes, a few metres in diameter, occasionally appeared, causing seepage.[13] Such occurrences were not widely publicised, presumably because no one had an interest in drawing attention to a potential problem that may affect both the extraction and the tourism industry around the Dead Sea. Eli Raz was one of the first to understand the link between the receding level of the lake and the formation of sinkholes. He warned that the sinkholes could soon endanger roads and tourist facilities along the entire western Dead Sea shore. However, like the mythological Cassandra or the biblical prophet Jeremiah, his predictions were ignored at first by local authorities.[14]

The problem has become worse over time as the lake's shoreline has continued to recede. The western shore of the Dead Sea has seen the highest number of sinkholes. The area most affected is the central part of the lake's historical shoreline, from Khirbet Mazin in the north to the western bank of the Lynch Strait in the south.[15] By the late 1990s, the sinkholes started to threaten the livelihood of Kibbutz Ein Gedi. In two separate incidents, people fell into newly formed holes: one in a campsite run by the kibbutz and another in a date palm grove. Luckily, no lives were lost. By the early 2000s, with the appearance of hundreds of sinkholes annually, the kibbutz felt under threat from potential lawsuits which could follow from injuries or damage to property. Hence, it was forced to abandon all its lakeside tourist facilities and several acres of date palms. In 2015, Route 90 had to be permanently diverted further inland through the Ein Gedi oasis because of the sinkhole threat close to the shoreline. 'In recent years nature is telling us something very clear that must not be ignored', local resident and history teacher Merav Ayalon told a reporter for *Haaretz*. 'Nature is telling us we created a problem which we must recognize with open eyes.'[16]

On the Jordanian side of the lake, the area most affected by sinkholes is in the vicinity of the Lisan Peninsula and Jordan's Potash City. A forebear of things to come was an access road that collapsed in October 1992 as a result of a 20-metre-deep sinkhole in an area where the lake had receded west of the Lisan Peninsula. In March 2000, part of a recently built dike protecting an evaporation pool north of Lisan collapsed because of a sinkhole, with 56 MCM of brine pouring out into the northern basin of the lake.[17] Sinkholes also appeared in the agricultural area of the village of Ghor al-Haditha, east of the Lisan Peninsula. Today, local farmers complain that the earth literally disappears under their feet and that they can no longer use tractors when working on their tomato fields.[18]

Since the early 2000s, aerial footage, often produced using drones, has increasingly exposed the extent of the Dead Sea sinkhole phenomenon. In some cases, it reveals how the sinkholes have destroyed buildings, roads and parking lots, such as those in the abandoned lakeside resort of Mineral Beach, south-east of the caves of Wadi Murabba'at. In other cases, the footage shows a moon-like topography dotted with craters. Some sinkholes have filled with water and have vegetation growing around them. The sinkholes vary in size, with the largest reaching a diameter of 25 metres. Estimates in 2023 put the overall number of sinkholes in the vicinity of the Dead Sea at around 8,000. In the areas most affected, a completely new landscape has emerged.

In addition to being extremely disruptive and dangerous, some of the sinkholes are undeniably beautiful. Those sinkholes that have filled with water tend to assume different shades and colours, ranging between yellow, green, red and almost black. The colours depend on the composition of the water that accumulates in the sinkhole and on which algae or bacteria begin to develop in it.

At present, visits to the 'land of the sinkholes' are completely unregulated and can be risky. For instance, drivers that tried to test the terrain have had their cars fall into sinkholes. Subsequently, they have had to rely on heavy vehicles whose preying owners were prepared to extract the fallen cars, but only for a substantial fee. There are those who call on authorities in Israel to create safe routes for visitors to see this unique phenomenon. They include Moshe Gilad, who reports on domestic tourism for *Haaretz* and has written extensively on the Dead Sea region, and Professor Nadav Lansky of the Geological Survey of Israel. The lake, says Gilad, is not going to be refilled anytime soon. He therefore suggests that clearly demarcated paths should be put in place to allow tourists to see the wondrous, moon-like scenery of the sinkholes. He says that there are safe ways of doing so and that the main obstacle so far is the unwillingness of insurance companies. Lansky and his colleagues at the Geological Survey point out that enabling tourists to pass safely through potentially dangerous terrain is not unheard of. Yellowstone Park in Wyoming, for instance, has a path going through an active geothermal area. Hence, solutions that mitigate the risks can be found. In addition to the sinkholes, making the area accessible would enable visitors to reach the lake itself, where the salt forms beautiful features on the waterline.[19]

A SITE OF NATURE, A SITE OF TOURISM

In early 2020, a report on *Kan* TV revealed to the public in Israel a breathtakingly beautiful canyon through which excess briny water flows from the artificial pools in the southern basin of the Dead Sea back to the remaining lake in the north. Using a drone, the report showed how, over the last few decades, the water had carved a meandering path through the dried-out Lynch Strait. The reporter, Oren Aharoni, compared the view to the Grand Canyon in the United States.[20] Calls to make the 'secret river' publicly accessible followed, although doing so would require a solution that bypasses potential sinkholes as well as an extensive landmine clearance effort: the canyon runs right along the Israeli–Jordanian border.

The new canyon in the Lynch Strait is a reminder that, even as it recedes and although its future is uncertain, the Dead Sea still offers sites of exquisite beauty. For centuries, the Dead Sea area was described as a desolate place, reeking of sulphur, with authors emphasising how the lake itself was devoid of life and how fish in the Jordan River tried their best to avoid it lest they perish. Since the medieval period, Christian and Muslim authors described the lake as cursed because of its association with the destruction of Sodom and Gomorrah. Yet, during the twentieth century, with

A SITE OF RUIN, BEAUTY, HEALTH AND HOPE

the advent of international tourism, the image of the Dead Sea was transformed. While the association with the biblical story of Sodom persisted, thousands and eventually hundreds of thousands began to flock to the lake to witness its unique beauty.

One strand of tourism began in the early decades of the twentieth century, when the Zionist movement placed an emphasis on encouraging the Jews of Mandate Palestine to know and tour the country. The Judean Desert, and above all Masada, became a frequent destination for trips by youth movements and paramilitary organisation members in the period between the two world wars. A tragedy befell a group of young Jewish hikers who camped in the Ein Gedi oasis in April 1942, having come down from Masada: a grenade the group carried with it for protection accidentally exploded near a campfire, killing eight of the hikers.[21]

Another problem that hikers faced was knowing which trail was safe to follow. This was especially true for the cliff south of Ein Feshkha, which, in the first half of the twentieth century, used to touch the water. Anyone travelling on land along the lake's western shore needed to climb or circumvent this obstacle, which proved difficult for hikers who were not sufficiently familiar with the region's terrain. To assist travellers in the area, the first marked hiker's trail in the country was established in autumn 1947. Hillel Birger, an engineer and topographer from Tel Aviv, was troubled by accidents that hikers in the Judean Desert had encountered. He therefore enlisted the help of a group of volunteers, and together they climbed from Ein Feshkha to the cliff of Ras Feshkha above it, marking a trail of about 3 kilometres. Conscious that the trail passed through essentially Bedouin territory, the trail-blazer Birger and his group used ladders to paint the markings high up on the rocks, where they would not be easily erased.[22]

While the Zionist movement and, from 1948, the State of Israel prioritised developing the country, populating it and maximising its agricultural capacity, the 1950s saw the emergence of voices calling for greater awareness to the protection of nature. In 1954, the Society for the Protection of Nature in Israel, which had been established by teachers, academics and nature enthusiasts, held its first gathering and started planning activities across the country. By the early 1960s, the Society began to operate a field school in the Ein Gedi oasis, which aimed to educate the public about the region's flora, fauna and landscape.

However, the Society felt that the field school on its own was not enough. In 1964, Azaria Alon, one of the founders of the Society, published an opinion piece in which he called for the Ein Gedi oasis to be recognised as a nature reserve. This, he argued, 'would ensure, for us and for those who come after us, the preservation of its special and unique character, its nature, rocks, antiques and the sight of its landscape'. Alon emphasised how, amidst the surrounding desert, the patch of green that is Ein Gedi can be seen all the way from Masada, 16 kilometres away. He also praised the ibex, with their long and curved horns, 'the most splendid and special of the animals of Ein Gedi'.[23]

The campaign of the Society was successful and, in spring 1968, the government announced the creation of a nature reserve covering part of the oasis and the wadis that descend into it. A nature reserve was also established further north, in the oasis of Ein Feshkha that had come under Israeli control as part of the conquests of the Six-Day War. To balance the protection of nature at Ein Feshkha with the desire of holidaymakers to bathe in the Dead Sea, the area was divided in two: one for recreation and the other where no bathing was allowed.[24] One group that did not benefit from the establishment of nature reserves in the oases on the western shore of the Dead Sea were the Bedouins. In the past, their caravans passed along the Dead Sea shore, with the oases providing water and shade for the camels and their riders. The twentieth century, with its industrial development, expansion of settlements, conflicts between peoples and between states and

A SITE OF RUIN, BEAUTY, HEALTH AND HOPE

even its efforts to protect nature, increasingly pushed the Bedouins and their traditional way of life from the shores of the lake.

Efforts to conserve nature took a bit longer to emerge on the eastern side of the Dead Sea. Nonetheless, in 1987, a large section of the Dead Sea shore and its hinterland was designated by the Jordanian Royal Society for the Conservation of Nature as the Mujib Biosphere Reserve. Covering an area of about 220 square kilometres, the reserve is one of the largest in Jordan. It includes Wadi Mujib, Wadi Hidan and Wadi Zarqa Ma'in, which flow down from the hill country between Madaba and Karak. Within the reserve are species of fauna that once roamed freely throughout the area around the Dead Sea: Nubian ibex, striped hyena, Syrian wolf, caracal cat, golden jackal, honey badger, Arabian leopard and many more.[25]

The flash floods that the area often sees during the rainy season have, at times, caused tragic consequences for those touring the area. In 2018 a flash flood swept away a group of school children and their teachers, who had come to Zarqa Ma'in to get 'in touch with nature'. After search and rescue operations, twenty-one were confirmed dead.[26] To prevent tragedies like this recurring, Jordanian authorities now prevent access to places prone to flash floods, like the trail up Wadi Mujib, during the winter months.

A SITE OF HEALTH

When I began my research for this book a few years ago, one person I spoke to, a talented singer who lives in the area, told me that repeated bathing in the salty water of the Dead Sea helped to rid her dreadlocks of headlice where all other products failed. The Dead Sea and some of the sites around it have been associated with health treatments since antiquity. According to Josephus, asphalt that was harvested from the lake was an important ingredient in several medications. He also tells of how, suffering from ill health at the end of his days, King Herod came to Callirrhoe on the eastern side of lake to bathe in its hot springs.

The rift system within which the Dead Sea sits has several active faults. The closer thermal springs are to a fault, the higher their temperature. The hottest springs in the area are those of Hammamat Ma'in, situated amid scenic waterfalls in Wadi Zarqa Ma'in, just over 4 kilometres east of the lake. The site benefits from some sixteen springs with a water temperature that ranges between 45°C and 61°C, making it Jordan's hottest. Further down towards the Dead Sea shore, there are also twenty-one springs at Zara, near where ancient Callirrhoe once stood.[27] With water rich in potassium, magnesium and calcium, the health benefits of these hot springs are often promoted by travel guides.

While the lavish hot spring resorts of Ma'in and Zara in Jordan attract tourists today, those on the western side of the lake have been less fortunate. A small number of visitors began to frequent the hot springs near Kibbutz Ein Gedi in the late 1950s. In the 1960s a makeshift facility was built at the site, and it was replaced in the early 1980s by a massive concrete spa.[28] However, the Ein Gedi hot spring spa has been one of the victims of the receding Dead Sea shoreline. Originally built near the shoreline, the spa building is now more than a kilometre away from the lake. The area around it is sinkhole-prone. Therefore, the spa was officially closed in 2021 and today stands deserted.

One health tourism trend that the Ein Gedi spa helped to popularise during the years of its activity was coating one's body with Dead Sea mud. The idea that the mineral- and salt-rich mud of the Dead Sea could help treat rheumatic diseases by assisting blood circulation and removing toxins from the body was already circulating among visitors in the area in the 1960s and 1970s.[29] The health benefits of the Dead Sea's mud dovetailed with the broader reputation of the lake's water as being able to treat various ailments. It was with an eye to the potential market for health tourism that Novomeysky and his associates wanted to establish the Kalia hotel in the early 1930s. Novomeysky noted how his sister, who suffered from ill health, found that frequent bathing in the Dead Sea helped her get back on her feet.[30]

It was only in the late 1960s, with the growth of international tourism, that plans got underway to turn Ein Bokek on the south-western shore of the lake into a major health tourism centre. Two hotels and a spa were already in operation in 1968, and more were under construction. The early 1980s saw large-scale investments by the Israeli government to further develop Ein Bokek. The site became popular with Scandinavian tourists. Patients with skin and respiratory illnesses from northern Europe were recommended a sojourn by the Dead Sea. Domestic health tourism was also encouraged. One Israeli reporter, David Moshiov, cited a local Bedouin belief that a woman who bathed in the hot springs of the Dead Sea would become younger and, should she be childless, would soon give birth to a boy. According to Moshiov, it was also claimed that Queen Cleopatra herself used to come to bathe in this area of the Dead Sea to restore her beauty (an assertion that, if it appeared in a student essay, I would immediately jot down next to it 'reference needed!').[31]

Of course, the lake's health benefits are not confined to hearsay and folklore alone. Thanks to its altitude of more than 400 metres below sea level, sunbathing by the Dead Sea has been shown to be beneficial for various skin ailments. Psoriasis patients, for instance, can enjoy the health benefits of sunbathing by the lake while their exposure to harmful ultraviolet radiation, particularly UVB, is reduced. Moreover, bathing in Dead Sea water also brings benefits to the skin of psoriasis patients.[32]

The Dead Sea's growing reputation for promoting good health eventually lent itself to an industry of skin-care products. Kibbutz Ein Gedi began to commercially distribute Dead Sea mud in the 1970s. In the following decade, it started to collaborate with three newer kibbutzim, which were established near the Dead Sea after the conquests of the Six-Day War, in producing cosmetic products on a larger scale. These were initially marketed under the brand Dead Sea Health Products, although the company was later rebranded and became known internationally as AHAVA ('love' in Hebrew).[33] The company uses a trademarked blend of minerals sourced directly from the Dead Sea in all of its skincare products.

In the 2000s, however, the location of AHAVA's headquarters and manufacturing centre at Mitzpe Shalem, a small settlement established in 1970 some 12 kilometres north of Ein Gedi, within the occupied West Bank, created a reputational issue for the company. Pro-Palestinian activists began a campaign calling on customers to boycott AHAVA's beauty products because these were produced in the occupied territories. 'Ahava uses Palestinian natural resources without the permission of or compensation to the Palestinians', campaigners wrote in 2009.[34] Regular protests were held outside an AHAVA store in central London in 2010 and 2011. Eventually, the branch was closed.[35] Nonetheless, AHAVA was still considered a good investment by the Chinese conglomerate Fosun that purchased it in 2016. In recent years, AHAVA transferred its manufacturing hub and visitor centre to Ein Gedi, which is within the pre-1967 borders of Israel.[36] Communities and businesses around the Dead Sea are deeply influenced by the broader state of relations between the peoples of the southern Levant.

A SITE OF PEACE

Visitors to the cemetery of Kibbutz Ein Gedi may come across an unusual plaque. As we have seen, in 1966 the area provided the backdrop for an aerial battle between Israeli and Jordanian aircraft. The wreck of the Jordanian Hunter that crashed near Ein Gedi was subsequently located in a survey by archaeologists Gideon

Hadas and Asaf Oron. What remained of the aircraft was handed over to Jordanian authorities in 2009. Following an initiative by Hadas, a plaque commemorating the pilot, Lieutenant Muwaffaq Salti, was put up in the cemetery of the kibbutz in 2021, even though at the time of his death he was an enemy combatant.[37]

Formal diplomatic relations between Israel and Jordan date back to the peace agreement that Prime Minister Yitzhak Rabin and King Hussein signed in October 1994. The atmosphere that enabled this agreement was created by the Oslo Accords between Israel and the PLO a year earlier and the subsequent creation of the Palestinian Authority. For a brief moment it seemed as if the intractable Arab–Israeli conflict was heading towards resolution. However, the cause of peace has suffered multiple setbacks since then. The agreement with Israel remains highly controversial and unpopular in Jordan, where support for the Palestinian cause is very strong. Nevertheless, for the Jordanian side of the Dead Sea, the peace agreement has led to very clear dividends in terms of economic development, tourism and archaeological research.

Prior to the peace agreement, the north-eastern side of the Dead Sea and the area near the Jordan River was a military zone. Due to its history of cross-border infiltrations, access to this area was restricted by Jordanian authorities. This changed after 1994. The peace agreement increased the volume of tourists and prompted investment in the hospitality sector. Soon, preparations were underway to establish a hotel complex on the north-eastern corner of the lake, near the village of Sweimeh (Suwayma). Before construction work began, archaeologists surveyed this area, where the main findings were ancient cemeteries, as well as Zara further south, where findings were much richer. As a result of this survey, plans for the hotel complex at Sweimeh were adapted slightly.[38] Today, there are more than a dozen luxurious hotels on the north-eastern shore, as well as public beaches on the lakeside further south.

In general, the second half of the 1990s saw a flurry of archaeological activity across the Kingdom of Jordan. For instance, in Madaba a new archaeological park was inaugurated and in Ghor es-Safi important work was done to uncover the remains of the Monastery of St Lot. From 1996 onwards, excavations north of the Dead Sea at the site of Bethany Beyond the Jordan, very near the border, revealed the remains of five churches and a baptistery dating back to the fifth and sixth centuries CE.[39] Since 2000, a number of new churches have been constructed at the site by different Christian denominations. One of these is the Greek Orthodox St John the Baptist Church of the Jordan River, built in 2003. Alongside the visitors' centre, plans are underway for a new Museum of Baptism for tourists and pilgrims visiting the eastern bank of the traditional baptism site. Bethany Beyond the Jordan, which has been recognised as a UNESCO heritage site, attracted some 200,000 visitors in 2022. The Jordanian government hopes to increase this number to 1 million visitors per year by 2030.[40] None of this would have been possible had the area remained a closed military zone, as it had been before the peace agreement of 1994.

A few dozen metres to the west, across the Jordan at Qasr al-Yahud, the picture is more complex. As we saw in the previous chapter, a few Christian churches were established on the western bank of the Jordan in the 1920s and 1930s. However, in the years following the Six-Day War, pilgrimage to the site ceased, as it became the scene for skirmishes between the IDF, Palestinian militants and Jordanian troops. Visits to the site by Christian pilgrims were renewed in the 1980s, after Israeli authorities allowed limited access to the banks of the Jordan.[41] But the area at large was still strewn with landmines, and the churches and monasteries near the river remained inaccessible.

Then, in the 2010s, the international charity organisation HALO contacted the Israeli Ministry of Defense with the suggestion to clear some of the minefields in the West Bank. Qasr

A SITE OF RUIN, BEAUTY, HEALTH AND HOPE

al-Yahud was selected. Remarkably, despite the at times strained relationships between all the relevant actors – the Israeli government, the Palestinian Authority, King Abdullah II of Jordan and the various Christian churches – all agreed for the mine-clearing operation to begin. In 2019, I had the privilege to visit the site as the mine-clearing project was progressing. Lasha Bluashvili, HALO's friendly Georgian team leader, showed me around those buildings that were already safe to enter. The Ethiopian monastery still had bullet marks on its external wall. One of the rooms of the Franciscan chapel of St John the Baptist had a bed with a mattress and pillow on it that looked as if they had been left there, undisturbed, since the building was evacuated in the late 1960s. The chapel has a balcony on its roof that offers a splendid view of the eastern side of the river and the new churches of Bethany Beyond the Jordan. The project was completed in spring 2020. In total, an estimated 4,000 mines were cleared, and the buildings were eventually returned to the respective churches. Despite a very challenging terrain in terms of topography as well as security and politics, the area around the Dead Sea continues to offer an incredible assortment of sites of historical and religious significance alongside its natural beauty. Given the right amount of human will, perhaps it can be preserved for future generations.

CAN THE DEAD SEA BE SAVED?

The northern basin of the Dead Sea is very deep. Estimates regarding how much more the level of the lake will continue to drop before the reduced inflow and evaporation reach a new equilibrium vary. There are a few unknowns, including to what extent climate change will affect the region and whether the amount of water removed by the extraction industry from the northern basin to evaporation pools in the south will change. Ibrahim Oroud of Mutah University in Karak calculated various scenarios for the lake's future level and suggested the new equilibrium could be between -543 metres, with the lake covering 480 square kilometres, and -585 metres, where the lake would shrink to an area of 435 square kilometres. Without any restoration measures and given expected population growth, global warming and subsequent pressures on freshwater resources, he predicts that the level of the Dead Sea will drop to the lower end.[42] In such a scenario, the salinity of the lake will be so high that it will likely become a brine pond.

There have been several attempts by activists, organisations and artists to draw international attention to the plight of the Dead Sea in the hope that action can be taken to reverse its demise. For instance, between 2011 and 2021, American artist Spencer Tunick took a series of photos of hundreds of naked men and women floating on and standing by the lake. For some of these images, he chose to cover the models in white paint to evoke the biblical story of Lot's wife.[43] Another initiative saw twenty-eight swimmers from nine countries swim together across the Dead Sea from east to west, from Jordan to Israel, in November 2016. This arduous crossing took seven hours to complete, and the swimmers had to wear special masks so that they didn't accidentally swallow any of the lake's deadly water. Another swim across the Dead Sea, this time near the estuary of the Jordan River, was subsequently completed by Yusuf Matari, a Palestinian, Munqeth Meyhar, a Jordanian, and Oded Rahav, an Israeli. Their joint effort, documented in the film *Dead Sea Guardians*, was a call for international cooperation to save the vanishing lake. 'We need to save the Dead Sea because it is a wonderful place,' says Rahav, 'it is a meeting place of all the civilisations of the Middle East. It is a place of healing, even though it is set in very challenging conditions. It is a wonder of nature.'[44]

Alas, the prospects of saving the lake do not look very promising. Over the last two decades, two proposals to bring more water to the Dead Sea have been put forward. Neither proposal considers refilling the Dead Sea back to its level in the 1950s or 1960s. Instead, the aim is to halt the decline and stabilise the lake's level by bringing an additional several hundred MCM to it each year.

The better known of the two proposals is the Red Sea–Dead Sea Water Conveyance. As we have seen, the idea of connecting these two bodies of water has been raised before. However, the environmental crisis in the Dead Sea rekindled interest in it. In 2005 the Jordanian, Israeli and Palestinian governments agreed to examine the feasibility of pumping water from the Red Sea near Aqaba and bringing it to the Dead Sea, thereby stopping the latter's decline. The World Bank agreed to coordinate donor funding for this project and to manage the various studies that would need to be undertaken. In its most ambitious iteration, the Conveyance was set to bring 1,200 MCM to the Dead Sea.[45] However, over the following years, the plan evolved. By 2017, the project's main objective shifted to increasing the supply of drinkable water in Jordan. Water scarcity in Jordan had become more severe, even though in the peace agreement of 1994 Israel had undertaken to supply Jordan with an annual quota of water from the Sea of Galilee. Hence, revised plans for the Red Sea–Dead Sea Water Conveyance envisaged desalinating up to 700 MCM of water in a specially built plant near Aqaba. The desalinated water would then be transported north through Jordanian territory using pipes. Most of the water would be used in the Amman area while the rest, along with the refuse brine from the desalination process, would pour into the Dead Sea near the Lisan Peninsula. Ultimately, the Dead Sea was to receive an additional 235 to 310 MCM.[46]

There were concerns about the costs associated with this project, as well as with the impact it would have on the environment in the Gulf of Aqaba and in the area through which the pipes would pass. There were also concerns about how the water introduced into the lake would affect the composition of the Dead Sea. The stratification and dilution of Dead Sea water by this new inflow could cause losses for the Dead Sea Works in Israel and the Arab Potash Company in Jordan. Leehee Goldenberg, an environmental lawyer and resource governance expert who successfully litigated against ICL's pumping of Dead Sea water, says jokingly that one of the few things that she and the extraction company can agree on is that they both oppose the Red–Dead project.[47] To allay such concerns and to better test the impact of introducing a new type of water to the Dead Sea, it was decided to begin with a pilot: pumping 300 MCM from the Red Sea, desalinating 65 MCM to provide freshwater for Aqaba and the southern part of Israel, and then conveying a mix of the refuse brine and the remaining seawater balance (235 MCM) to the Dead Sea. However, as of 2025, the Jordanian plan is to construct only an Aqaba-Amman Water Desalination and Conveyance Project, and the idea of bringing water from the Red Sea to the Dead Sea has been shelved.[48]

EcoPeace, an environmental organisation that is run jointly by Israelis, Jordanians and Palestinians, put forward an alternative solution: increasing the amount of water that flows down the Jordan River to the Dead Sea, and limiting the amount of water used by the mineral extraction industry in Israel and Jordan. The latter could be achieved by developing more efficient ways of using a smaller amount of water drawn from the lake's northern basin. 'If the mineral industries, on both sides, were to pay for the Dead Sea water they use', EcoPeace argues, the companies 'would develop technologies that could reduce their Dead Sea water consumption'. Increasing the amount of water reaching the Dead Sea would involve allowing both treated wastewater and desalinated seawater from the Mediterranean to flow through the Jordan River. EcoPeace believes that, with proper infrastructure, it would be possible to make available between 300 and 400 MCM of treated wastewater for this purpose. They also suggest compensating farmers in the area for any water they would miss out on as a result.[49]

One advantage of this solution is that it would not only help to stabilise the Dead Sea but also contribute towards reviving the Jordan River. Over the last few decades, the river has seen not only a massive reduction in the amount of water flowing through it but also a severe drop in the water's quality. At present, the Jordan receives treated and untreated wastewater. It also has high levels of salinity. This is because there are saltwater springs west of the Sea of Galilee, a legacy of the Pleistocene Lake Lisan. Since the 1960s, water from these springs is gathered by a canal and is poured into the Jordan River south of the Degania Dam so as to protect water quality in the Sea of Galilee. As a result, the biodiversity on the banks of the Jordan and inside the river itself has been severely depleted.

In recent years some tentative steps were taken to make the possibility of allowing a larger amount of water to flow down the Jordan River more tangible. In terms of diplomacy, in November 2022 the Israeli and Jordanian governments agreed to work together to clean up the Jordan River. An official statement recognised that 'Rehabilitation of the Jordan River is an important goal for both countries due to its rich historical heritage, its importance as a main tourism site, as well as its high ecological value'.[50] There are even some, though at this stage very limited, efforts to realise this declaration. Israel's Water Authority is currently working on a project that would see it release more water, and of better quality, down a stretch of the Jordan River, although that water would be captured a few kilometres downstream and diverted towards agricultural use.

Another important development is related to desalination. Much of the water used in Israel today comes from five desalination plants on the Mediterranean coast. The availability of desalinated water has enabled Israel to increase the amount of water it supplies each year from the Sea of Galilee to the Kingdom of Jordan. To meet this obligation, as well as to ensure that the level and quality of the water in the Sea of Galilee are preserved, Israel has recently begun to divert some desalinated water into this lake. Israelis call this water management project the 'Reverse National Water Carrier': unlike the one constructed in the 1960s, which took water away from the Jordan Valley basin, this new project brings external water into it. At present, every MCM of water is counted and there is none to spare. However, perhaps in the future it would be possible to bring larger amounts of water into the Sea of Galilee so that the Degania Dam could be opened more regularly, allowing more water to flow down the Jordan River. Some of it might be captured by farmers downstream. But, if the diplomatic atmosphere in the region improves after the nadir of 2023–2025, perhaps Israelis, Jordanians and Palestinians can agree to ensure that a set amount of water reaches the Dead Sea every year. Some years ago, Eli Raz calculated that, given the reduced size of the lake, roughly 850 MCM annually would suffice to stabilise its level.[51] Because the surface area of the lake has continued to decrease, stabilisation could be reached with even a smaller amount of water.

Increasing desalination in the Mediterranean Sea would not come without environmental consequences. For starters, it would increase the amount of brine waste that is produced in the process and discharged off the Mediterranean coast. Furthermore, operating desalination plants requires a good deal of energy. But, at the very least, this idea deserves further consideration and detailed study. For instance, it would be worth exploring whether a new desalination plant could be powered by solar energy.[52] As Leehee Goldenberg points out, 'the Dead Sea is a regional symbol. It is also a symbol of humanity. In addition to its beauty, it has historical value. Our generation needs to leave behind a legacy. Regionally, historically and in terms of industry, the Dead Sea is a legacy worthy of preservation.'

EPILOGUE: A HITCHHIKER'S GUIDE TO THE BELEAGUERED DEAD SEA

WHEN I STARTED working on this book, my plan was to finish on an optimistic note, emphasising the beauty of the Dead Sea area and discussing plans to save it. However, 2023 to 2025 have not been good years for optimism in this region of the world. As I finish writing, the tragic war which began on 7 October 2023 is still raging. Though sufficiently far from Gaza, the Dead Sea region and some of the people mentioned in previous chapters are not unaffected. Along with other archaeologists, Dr Joe Uziel, who heads the Dead Sea Scroll unit of the Israel Antiquities Authority, was called to help identify bone fragments and human remains in the kibbutzim of the western Negev that were attacked on the first day of the conflict.[1] Foreign tourists disappeared from the hotels on the western side of the Dead Sea. Instead, these hotels began hosting families that were evacuated from the kibbutzim on the front line. In April 2024, hundreds of Iranian missiles were fired on Israel, only to be intercepted by an international coalition that included the United States, Israel and Jordan. At least one intercepted missile fell into the Dead Sea.[2] Soon, images appeared on social media showing the Iranian missile 'doing the Dead Sea float'. In a subsequent Iranian attack in

October of that year a Palestinian worker from Gaza was killed near Jericho by rocket debris that fell from the sky.

Let me therefore conclude with a quick tour around the lake and an appraisal, at the same time pessimistic and optimistic, of some of the things that new visitors to the area might expect to find. Tell es-Sultan, one of the oldest permanent settlements in the world, the site of Elisha's Spring, was almost empty on a bright winter morning in December 2022. One couldn't help thinking more people should come to see the round, stone-built tower that was exposed during Kathleen Kenyon's excavations and dates back to the Pre-Pottery Neolithic. It is likely the oldest tower in the world still standing. In 2023 Tell es-Sultan was recognised by UNESCO as a World Heritage Site, despite objections from some Israeli officials and right-wing politicians.[3] Few archaeological sites in the world are more deserving of this title. Indeed, Jericho has much to offer. Within a radius of a few kilometres, one is able to visit Mount of Temptation (accessible by cable car from Tell es-Sultan), Hisham's Palace, where a dome, funded by the Japanese Ministry of Foreign Affairs, now covers the large Umayyad mosaic, and the tree of Zacchaeus.

Jericho's Shalom al Israel synagogue has an interesting story to tell. In 1936, a wealthy Palestinian from Jerusalem, Husni Shahwan, uncovered the mosaic of this sixth- or early-seventh-century synagogue while building a house on the outskirts of Jericho. He took care to construct the basement of his house in such a way that would not destroy the mosaic. In 1967, after the Six-Day War and the Israeli conquest of the area, Defense Minister Moshe Dayan thanked the Shahwan family for preserving this relic.[4] However, gaining access to the remains of the synagogue today may prove difficult because of political tensions.

The sign over Jericho's city-centre police station bears three words that remain incredibly contentious at the time of writing: 'State of Palestine'. The Oslo Accords of the 1990s gave birth to the Palestinian Authority, an autonomous governmental entity

EPILOGUE

that had some but not all of the characteristics of an independent state. The municipality of Jericho and part of the Gaza Strip were the first territories to come under the control of the Palestinian Authority in 1994, as part of what was meant to be an interim phase on the road to a more permanent Israeli–Palestinian settlement. However, as the peace process broke down, Jericho, like the rest of the West Bank, has remained suspended in this interim phase for the last thirty years.

One soon leaves the small territory administered by the Palestinian Authority when travelling north-west from Jericho towards the spring of al-Auja. Our local guide, Osama, who defines himself as a peaceful freedom fighter, showed us the increasing presence of Jewish settlements in this part of the Jordan Valley. Yitav, an agricultural settlement, has been in place since the 1970s. Mevo'ot Yerikho, much nearer to Jericho, was established in 1999 but only received formal recognition from Israeli authorities in 2019.[5] In recent years yet another Jewish settlement, the name of which still does not appear on maps, popped up between Mevo'ot Yerikho and Yitav. Since 2020, there has been a growing number of violent incidents between Jewish settlers and Palestinians in the Jordan Valley. Wadi Auja, further west, is scenically beautiful, but I would not recommend visiting the area under the present political climate.

In early 2023, pent-up tensions turned into open violence south of Jericho. Palestinian militants carried out a number of attacks in the vicinity of the Almog Junction, where a petrol station and several stalls cater for tourists driving down from Jerusalem to the Dead Sea. A US citizen visiting the area was killed in one of these attacks. Concurrently, the IDF carried out raids targeting a Hamas cell in the former refugee camp of Aqbat Jabr. As a result, eight Palestinians were killed. If things weren't bad enough, several acres of date palm groves belonging to Kibbutz Almog, a settlement established in 1979, were soon set on fire. *Haaretz* journalist Nir Hasson wrote a poignant piece about the violence in the area. He pointed out that it wasn't clear whether the arsonists were Palestinians retaliating against Israeli incursions or right-wing Jewish Israelis seeking to get back at the kibbutz because some of its members offered to help a father from Aqbat Jabr whose teenage son had been killed.[6] The northwestern corner of the Dead Sea has known far better times in the past. One hopes better times also lie ahead.

Other parts of the Dead Sea shore may feel very far removed from tension and violence. On the Jordanian side of the lake, visitors who are willing to venture beyond the tourist resorts near Sweimeh and head south on Route 65 may find reward in the small but very informative Museum at the Lowest Place on Earth in Ghor es-Safi. The museum explains the history of human settlement in the area and displays several artefacts that were uncovered in local excavations. From the museum, if the weather is not too hot, one can climb up to see the remains of the Monastery of St Lot that was built at the entrance to a grotto in the Byzantine period. In addition to archaeological remains, the site offers an excellent vantage point on the fertile plain south of the Dead Sea. Should the weather be too hot, one could head back north for a trip upstream, through the narrow ravine of Wadi Mujib, which may offer a good way to cool down.

The springs of Zara further north sustain lush vegetation east of Route 65, where one can find the hot spring spa and a few other tourist resorts that have been established in recent decades. West of the road, above the lake, are the archaeological remains of ancient Callirrhoe. These were excavated in the late 1990s and the area was earmarked to become Az-Zara Archaeological Park.[7] Alas, today, because of the hazards created by the retreating Dead Sea shoreline, the site remains fenced off and inaccessible.

On the Israeli side of the lake, Kibbutz Ein Gedi's botanical garden offers some escape from the heat, as well as a sharp contrast with the surrounding desert environment. The botanical garden

EPILOGUE

covers most of the area of the kibbutz, where the community lives among several hundred species of plants. Some of these, like date palms and the so-called apple of Sodom (*Calotropis procera*), can be found elsewhere around the lake, but many others were imported from several countries around the world. The tall baobabs, planted in the early 1960s, are particularly impressive. The garden also features a specimen of *Commiphora gileadensis*, a bush that is sometimes associated with the legendary balsam of antiquity. In addition to the large array of plants, some of the ancient anchors that were found on the shore of the Dead Sea are on display near the communal dining hall. Ibex from the nearby Ein Gedi nature reserve sometimes hang out on the ridge below the northern side of the kibbutz.

A few kilometres south of the kibbutz, on the left-hand side of Route 90, stands the abandoned Ein Gedi hot spring spa building. In a conversation I had with *Haaretz* reporter Moshe Gilad, he proposed turning the former spa into an art gallery. He pointed out that it would be ideally suited for displaying the works of the artist Sigalit Landau, who has been using the Dead Sea as a focal point for her art over the last twenty years. In one of her most famous works, *Salt Bride*, Landau submerged a black dress in the lake for several weeks and documented the process of it accumulating salt crystals until it became completely white and glimmering.[8]

The cluster of about a dozen hotels in the Ein Bokek resort on the western shore of the southern basin has a capacity of approximately 4,000 rooms. This tourist complex has the opposite problem to the one faced by the lakeside resorts on the shores of the northern basin of the Dead Sea. The beaches near Kalia in the north-west and the hotels near Sweimeh on the north-eastern side of the lake have to constantly adapt to the falling level of the Dead Sea. Resorts in the north need to 'chase' the waterline, which drops by about a metre per year, periodically moving their facilities lower and extending their access paths to reach the shoreline. Conversely, guests staying at Ein Bokek in the south benefit from sandy beaches and from a more or less steady water level because the hotel area sits next to one of the huge artificial evaporation pools operated by ICL. In fact, the hotels have had to put pressure on ICL to make sure the water level does not rise any higher as huge quantities of salt accumulate at the bottom of the evaporation pool. In 2022, Israel's Environmental Ministry called on ICL to employ a dredger regularly to scrape salt from the bottom of the pool next to the hotels at Ein Bokek to prevent them from overflowing.[9] The scraping process creates a new dilemma: what should be done with the massive amount of salt collected by the dredger? ICL is in the process of coming up with a comprehensive plan.[10]

On the edge of the evaporation pools south of Ein Bokek's sandy beach area, visitors can find the so called 'salt mushrooms', mushroom-like structures composed of salt crystals that form thanks to the combined work of waves and evaporation. These features are proving very popular among selfie-takers. A few kilometres away, at the foot of Mount Sodom, ICL recently opened a museum and visitor centre, presenting the past and present of extraction at the site in a positive light. The museum gives a sense of working and living conditions at the southern end of the Dead Sea in the 1930s and 1940s, when Moshe Novomeysky was still at the helm.

When driving further south along Route 90, past the huge contemporary industrial complex of the Dead Sea Works, one of the most surreal scenes of the Dead Sea area is revealed. In the nineteenth century, part of this area was covered with marshes, where water from the wadis to the south and south-west accumulated. 'The place positively swarmed with birds in countless myriads, rising at every step', wrote Tristram in the 1860s.[11] The marshes were gradually replaced by evaporation pools and by the agricultural lands of two Israeli villages, Ein Tamar and Neot HaKikar. ICL also uses the area for intense quarrying to supply gravel for the construction and maintenance of their dikes. Over

EPILOGUE

time, some of the defunct quarries filled up with a mixture of groundwater and runoff, producing brackish ponds around which thickets of reeds grow. The ponds were given the name Milhat Sdom. They attract several species of ducks and birds and may be of interest to birdwatchers. The place is surreal, because the beauty of the ponds stands in sharp contrast to the quarrying and the constant movement of trucks that continues right next to them.

ICL's extraction concession in the Dead Sea area is up for renewal in 2030. Environmental activists hope that, as part of the new concession, the company will be required to take more stringent measures to safeguard the environment near the Dead Sea, for instance by preventing spillage of industrial waste and pollution into nearby wadis.[12] As we have seen, the organisation EcoPeace is calling on governments to make ICL and its counterpart across the lake in Jordan, Arab Potash Company, pay for the water they draw from the northern basin of the Dead Sea. They argue that introducing such payments would encourage the companies to draw less water and use it more economically, thereby slowing the rate of the lake's decline. The retreating shoreline is already a problem for tourist resorts near the Dead Sea. Sooner or later, the continued shrinking of the lake will become a serious problem for the extraction industry too.

This is where the story of the Dead Sea currently stands. Though a small lake, its significance resonates in many different cultures beyond the region itself. I hope that readers have come to appreciate how important the lake has been historically, and that they will be mobilised to care about its continued survival. For its future to be brighter than the present, a change in direction is needed, and such a change requires international cooperation and peace. The Dead Sea won't be saved without them.

ACKNOWLEDGEMENTS

WRITING A BOOK that seeks to cover a period of approximately 10,000 years meant going on a long trip outside of my research comfort zone in modern international history. Luckily, many friends and colleagues were generous with their time and willing to help me along the way.

I learned a great deal about geology and paleoclimatology from Daniel Palchan, Daniel Hill and Graham McLeod. I would like to thank Lorenzo Nigro, Gideon Hadas and Phillip Silvia for sharing some of their insights on the archaeology of the Dead Sea region. I am also grateful to James Tabor for his advice on the Dead Sea Scrolls.

I am heavily indebted to Alan Murray, Joanna Phillips, James Doherty and Beth Spacey for their advice and assistance in locating sources on the Frankish period. Harry Munt kindly helped me to identify relevant medieval Arabic geographers, and Sara Alkahtani produced excellent translations of some of their texts. I would also like to thank Maroula Perisanidi for her assistance with Greek translations and Rebecca Darley for her advice on the Madaba map. At times when accessing sources was difficult, Mike Belantara, Tzachi Ein Gil and Sarah Irving were willing to help out, and for that I am grateful.

ACKNOWLEDGEMENTS

I would like to express my gratitude to Leehee Goldenberg, Oded Rahav, Eli Raz, Moshe Gilad, Gidon Bromberg and Yana Abu Taleb, not only for sharing with me their thoughts and experiences, but also for their continued efforts to save the Dead Sea.

Researching the history of the Dead Sea required extensive travel. I would like to thank the Gerda Henkel Foundation for funding my initial fieldwork in Ein Gedi. Rina Yaaran, the archivist of Kibbutz Ein Gedi, and local historian Neri Erely both helped me a great deal with my research. I am very grateful to my friend Osama for hosting me in Jericho, to HALO's Louise Vaughan, Ronen Shimoni and Lasha Bluashvili for arranging my visit at the mine clearance project in Qasr al-Yahud, and to Carol Palmer and Firas Bqa'in for all their assistance during my research at the CBRL in Amman. Kariman Mango and Jack Green gave me excellent advice when planning my tour of archaeological sites in Jordan, and Mohammed Shapsoug spared no effort in helping me visit these. Felicity Cobbing and the staff at the PEF Archive at Greenwich have been immensely helpful throughout the years I have been working on the history of the Dead Sea. Elaine Solowey showed me 'Methuselah' and the other old-new date palms at the Arava Institute for Environmental Studies and kindly let me taste one of their exquisite dates. Sarah Sallon taught me a great deal about the process of bringing Roman-era date seeds back to life. I thoroughly enjoyed working with Maor Kohn and Ofer Nisim on the history of water management in the Jordan Valley. I would also like to thank Alrun Gutow and Kat Streicher for enabling me to obtain images from the Vorderasiatisches Museum in Berlin, and my Head of School, Sanjoy Bhattacharya, who helped me secure the time needed to complete this project.

The text of this book is much richer than it otherwise would have been thanks to the thoughtful suggestions of all those who commented on earlier versions or parts of the manuscript: Kimberley Thomas, Francesca Morphakis, Gali Jaffe, Johanna

ACKNOWLEDGEMENTS

Stiebert, Oren Ostersetzer, Matias Sametband, Maddie Shaw, Alex Worsfold, Anne Caldwell and, last but not least, my parents.

I would like to thank my editor at Yale University Press, Jo Godfrey, for her endless patience and for being a constant source of wise and sensible advice. I am also grateful to Katie Urquhart and her team for their excellent work on the book's cover, maps and images.

This is a good opportunity to thank (and apologise to) my amazing kids, Nima and Adri, who have had to visit the Dead Sea and hear about it far more than they cared to (Adri especially is not a fan). My heartfelt thanks also go to my Vanja, for everything, and for taking many of the photos that accompany this book. My talented nieces, Adi Biran and Danielle Biran Angel, did an excellent job editing the video showing floodwater flowing into and spreading across the Dead Sea. Have a look at the video on the book's webpage.

Sadly, one of the first people with whom I discussed this project several years ago, Ronnie Ellenblum, passed away in 2021. Ronnie, one of the kindest and most innovative historians I have known, supervised my undergraduate dissertation on nineteenth-century archaeological exploration in Palestine. I dedicate this book to his memory.

TIMELINE

c.135,000–75,000 years ago	Lake Samra
c.70,000–15,000 years ago	Lake Lisan
c.15,000–12,000 BCE	decline of Lake Lisan
c.11,000 BCE	emergence of the Natufian culture
c.10,800–9,700 BCE	Younger Dryas climatic event
early tenth–early ninth millennium BCE	Pre-Pottery Neolithic A (PPNA)
mid-ninth–mid-seventh millennium BCE	Pre-Pottery Neolithic B (PPNB)
c.6200 BCE	collapse of an ice sheet in North America; period of cold climate
c.4600–3600 BCE	Late Chalcolithic period and Ghassulian culture
c.3300–3000 BCE	Early Bronze Age I
c.3000–2700 BCE	Early Bronze Age II
c.2700–2350 BCE	Early Bronze Age III
c.2000–1600 BCE	Middle Bronze Age

TIMELINE

c.1650 BCE	destruction of Tall el-Hammam
c.1600–1200 BCE	Late Bronze Age
tenth century BCE	raid by Egyptian Pharaoh Sheshonq I
c.840 BCE	Moabite Stone (Mesha Stele)
c.798–769 BCE	reign of King Amaziah of Judah
c.759 BCE	earthquake or series of earthquakes
539 BCE	Cyrus the Great defeated the Babylonians
332 BCE	region conquered by Alexander the Great; revolt in the city of Samaria
c.311 BCE	naval battle on the Dead Sea
284–246 BCE	Ptolemy II Philadelphus; Zenon papyri
176–170 BCE	Ptolemy VI of Egypt
169–168 BCE	Antiochus IV attacked Jerusalem
167 BCE	start of Hasmonean revolt
135 BCE	murder of Simon Thassi at Dok
104–76 BCE	reign of Alexander Jannaeus
40 BCE	Hasmonean Kingdom overrun by the Parthians
37 BCE	Herod's rise to power
31 BCE	Battle of Actium; death of Cleopatra; earthquake in the Dead Sea area
4 BCE	death of Herod
66 CE	outbreak of the First Jewish–Roman War
c.68 CE	Dead Sea Scrolls hidden near Qumran

TIMELINE

70 CE	destruction of the Temple in Jerusalem
73/74 CE	siege of Masada
132–135	Bar Kokhba Revolt
330	Constantine I moved the capital of the Roman Empire to Constantinople
527–565	reign of Emperor Justinian I
614	the Byzantine southern Levant was overrun by the Persian-Sasanian Empire
628/629	Byzantine rule briefly restored by Emperor Heraclius
632	death of the Prophet Muhammad
661	Umayyad Caliphate established
724–743	reign of Caliph Hisham ibn Abd al-Malik
749	major earthquake in the Dead Sea area
750	Umayyad Caliphate overthrown by the Abbasids
1099	Frankish Kingdom of Jerusalem established
1187	Battle of Hattin: Franks defeated by Saladin
1260	Battle of Ain Jalut: Mongols defeated by the Mamluks
1269	construction of the *maqam* at Nabi Musa by Baybars
1516	Ottoman conquest
1520–1566	reign of Sultan Suleiman the Magnificent

TIMELINE

1865	establishment of the Palestine Exploration Fund (PEF)
1869	opening of the Suez Canal
1900	measurements begin at the PEF Rock
1902	*Altneuland* published by Herzl
1917–1918	British conquest of Palestine
1927	major earthquake in the Dead Sea area
1930	beginning of Palestine Potash Company concession
1946/1947	discovery of the Dead Sea Scrolls
1947	UN Partition Plan
1948	end of British Mandate; first Arab–Israeli War
1952	potash extraction in Israel nationalised; establishment of Dead Sea Works
1961	first part of the East Ghor Canal inaugurated
1964	completion of Israel's National Water Carrier
1967	Six-Day War
1968–1970	intermittent fighting in the Jordan Valley
1977	southern and northern basins of the Dead Sea separated owing to the falling water level; construction of Arab Potash Company plant in Jordan begins

TIMELINE

1994	start of Palestinian Authority control over Jericho; Peace treaty between Jordan and Israel
1996	beginning of excavations at Bethany Beyond the Jordan
2020	completion of landmine removal at Qasr al-Yahud
2023–2025	Israel–Gaza War

NOTES

INTRODUCTION: A PLACE OF MANY CONTRADICTIONS

1. Aristotle, *Meteorologica*, vol. II (Cambridge, MA: Harvard University Press, 1952), 159.
2. Fernand Braudel, *The Mediterranean and the Mediterranean World in the Age of Philip II* (London: Collins, 1972).
3. David Abulafia, *The Great Sea: A Human History of the Mediterranean* (London: Penguin, 2012), xxvi–xxxi.
4. K.N. Chaudhuri, *Trade and Civilisation in the Indian Ocean: An Economic History from the Rise of Islam to 1750* (Cambridge University Press, 1985); Marcus Rediker, *Outlaws of the Atlantic: Sailors, Pirates, and Motley Crews in the Age of Sail* (Boston: Beacon Press, 2014); David Abulafia, *The Boundless Sea: A Human History of the Oceans* (London: Allen Lane, 2019).

CHAPTER 1 A HARSH BUT WELCOMING SITE

1. Pausanias, *Description of Greece* (London: Bell, 1912), 314.
2. Ellsworth Huntington, *Palestine and its Transformation* (Boston and New York: Houghton Mifflin, 1911), 198.
3. Wolfgang Zwickel, 'The Dead Sea in the Bible', in Martin Peilstöcker and Sabine Wolfram (eds), *Life at the Dead Sea: Proceedings of the International Conference held at the State Museum of Archaeology Chemnitz (smac), February 21–24, 2018, Chemnitz* (Münster: Zaphon, 2019), 19.
4. Andrea Orendi, 'Development and Importance of Agrarian Resources in the Dead Sea Region in the Bronze and Iron Age', in Peilstöcker and Wolfram, *Life at the Dead Sea*, 126.
5. Dorrik Stow, *Vanished Ocean: How Tethys Shaped the World* (Oxford University Press, 2012), viii; Thomas Halliday, *Otherlands: A World in the Making* (London: Penguin, 2023), 42.

6. Aharon Horowitz et al., *The Jordan Rift Valley* (Lisse: Balkema, 2001), 467.
7. Albert Mathieson Quennell, 'The Structural and Geomorphic Evolution of the Dead Sea Rift', *Quarterly Journal of the Geological Society of London*, 64 (1958), 3; Raphael Freund, 'A Model of the Structural Development of Israel and Adjacent Areas since Upper Cretaceous Times', *Geological Magazine*, 102:3 (1965), 189–205.
8. Zvi Garfunkel, 'Ha-tektonika shel transform yam ha-melach [The Tectonics of the Dead Sea Transform]', *Melach Haaretz*, 2 (2006), 4–10.
9. Ronnie Ellenblum et al., 'Crusader Castle Torn Apart by Earthquake at Dawn, 20 May 1202', *Geology*, 26:4 (1998), 303–6.
10. Horowitz, *The Jordan Rift Valley*, 498, 507, 513.
11. Klaus Bandel, Ikhlas Alhejoj and Elias Salameh, 'Geologic Evolution of the Tertiary-Quaternary Jordan Valley with Introduction of the Bakura Formation', *Freiberger Forschungshefte*, 23 (2016), 103–4.
12. Eli Raz, *Sefer Yam Ha-Melah* [Book of the Dead Sea] (Jerusalem: Nature and Parks Authority Press, 1993), 27; R. Weinberger et al., 'Quaternary Rise of the Sedom Diapir, Dead Sea Basin', *Special Paper of the Geological Society of America*, 401 (206), 33–51.
13. Jean Marie Rouchy and Antonio Caruso, 'The Messinian Salinity Crisis in the Mediterranean Basin: A Reassessment of the Data and an Integrated Scenario', *Sedimentary Geology*, 188–9 (2006), 35–67.
14. Bandel, Alhejoj and Salameh, 'Geologic Evolution', 122; Nicolas Waldmann et al., 'Stratigraphy, Depositional Environments and Level Reconstruction of the Last Interglacial Lake Samra in the Dead Sea Basin', *Quaternary Research*, 72 (2009), 1–15.
15. Louis Lartet, *Exploration géologique de la mer Morte de la Palestine et de l'Idumée* (Paris: Bertrand, 1877), 263–73.
16. Yuval Bartov et al., 'Lake Levels and Sequence Stratigraphy of Lake Lisan, the Late Pleistocene Precursor of the Dead Sea', *Quaternary Research*, 57:1 (2002), 9–21; Shahrazad Abu Ghazleh, 'Lake Lisan and the Dead Sea: Their Level Changes and the Geomorphology of their Terraces' (PhD thesis, Technische Universität Darmstadt, 2011), 8.
17. Mordechai Stein et al., 'Abrupt Aridities and Salt Deposition in the Post-Glacial Dead Sea and their North Atlantic Connection', *Quaternary Science Reviews*, 29 (2010), 567–75.
18. Eva Kaptijn, *Life on the Watershed* (Leiden: Sidestone Press, 2009), 15.
19. Horowitz, *The Jordan Rift Valley*, 461.
20. Yuval Bartov et al., 'Late Quaternary Faulting and Subsidence in the Central Dead Sea Basin', *Israel Journal of Earth Sciences*, 55 (2006), 17–31.
21. Naama Goren-Inbar et al., 'Evidence of Hominin Control of Fire at Gesher Benot Ya'aqov, Israel', *Science*, 304 (2004), 725–7; Irit Zohar et al., 'Evidence for the Cooking of Fish 780,000 Years Ago at Gesher Benot Ya'aqov, Israel', *Nature Ecology & Evolution*, 6 (2022), 2016–28.
22. Milutin Milankovitch (Milanković) suggested that there are separate cycles in the shape of Earth's orbit, the angle at which the Earth tilts, and the direction in which our planet's axis of rotation is pointed. These changes affect the Earth's position relative to the sun and therefore contribute to long-term changes in climate.

NOTES TO PP. 17-24

23. Ofer Bar-Yosef, 'The Natufian Culture in the Levant, Threshold to the Origins of Agriculture', *Evolutionary Anthropology*, 6:5 (1998), 159; Dorothy A.E. Garrod, 'A New Mesolithic Industry: The Natufian of Palestine', *The Journal of the Royal Anthropological Institute of Great Britain and Ireland*, 62 (1932), 257–69.
24. Bar-Yosef, 'The Natufian Culture', 159.
25. Anna Belfer-Cohen, 'The Natufian in the Levant', *Annual Review of Anthropology*, 20 (1991), 173.
26. Ibid., 171–8.
27. Anna Belfer-Cohen and A. Nigel Goring-Morris, 'Breaking the Mold: Phases and Facies in the Natufian of the Mediterranean Zone', in Ofer Bar-Yosef and François R. Valla (eds), *Natufian Foragers in the Levant* (Ann Arbor, MI: Berghahn, 2013), 544–61.
28. Phillip C. Edwards, 'Natufian Interactions along the Jordan Valley', *Palestine Exploration Quarterly*, 147:4 (2015), 272–82.
29. Yehuda Levi et al., 'Harnessing Paleohydrologic Modeling to Solve a Prehistoric Mystery', *Scientific Reports*, 9 (2019).
30. Ofer Bar-Yosef, 'The Walls of Jericho: An Alternative Interpretation', *Current Anthropology*, 27:2 (1986), 157–62.
31. Steven Mithen, *After the Ice: A Global History 20,000–5,000 BC* (London: Weidenfeld & Nicolson, 2003), 57.
32. Ibid., 48, 57.
33. For the coining of the phrase 'Neolithic Revolution', see: V. Gordon Childe, *Man Makes Himself* (London: Watts & Co., 1936), 39–40, 49–50, 74.
34. Gaia Ripepi, 'Mudbricks and Modular Architecture at Tell es-Sultan from the Neolithic to the Bronze Age', in Rachael Thyrza Sparks et al. (eds), *Digging Up Jericho: Past, Present and Future* (Oxford: Archaeopress, 2020), 215–16; Lorenzo Nigro, 'Jericho and the Dead Sea: Life and Resilience', in Peilstöcker and Wolfram, *Life at the Dead Sea*, 139.
35. Kathleen M. Kenyon, 'Excavations at Jericho', *The Journal of the Royal Anthropological Institute of Great Britain and Ireland*, 84:1/2 (1954), 105.
36. Nigro, 'Jericho and the Dead Sea', 140; Mithen, *After the Ice*, 67–8; Kenyon, 'Excavations at Jericho', 107.
37. Lorenzo Nigro, 'The Italian-Palestinian Expedition to Tell es-Sultan, Ancient Jericho (1997–2015): Archaeology and Valorisation of Material and Immaterial Heritage', in Sparks et al., *Digging Up Jericho*, 179; Mithen, *After the Ice*, 59.
38. Bar-Yosef, 'The Walls of Jericho', 158.
39. Ran Barkai and Roy Liran, 'Midsummer Sunset at Neolithic Jericho', *Time and Mind*, 1:3 (2008), 273–84; Nigro, 'The Italian-Palestinian Expedition', 179–80.
40. Mithen, *After the Ice*, 67.
41. Nigro, 'The Italian-Palestinian Expedition', 199–200.
42. Ghattas Sayej, 'A New Pre-Pottery Neolithic A Cultural Region in Jordan: The Dead Sea Basin', in *Australians Uncovering Ancient Jordan: Fifty Years of Middle Eastern Archaeology* (Sydney: Research Institute for Humanities and Social Sciences, University of Sydney; Amman: Department of

Antiquities, Jordan, 2001), 225–6; Phillip C. Edwards and Armin Schmidt, 'A Geophysical Survey at Zahrat adh-Dhra' 2 and its Implications for Pre-Pottery Neolithic A Architectural Traditions in the Southern Levant', *Jordan Journal for History and Archaeology*, 15:1 (2021), 108–9.

43. Hamoudi Khalaily and Jacob Vardi, 'The New Excavations at Motza: An Architectural Perspective on a Neolithic "Megasite" in the Judean Hills', in Hamoudi Khalaily et al. (eds), *The Mega Project at Motza (Moẓa): The Neolithic and Later Occupations up to the 20th Century* (Jerusalem: Israel Antiquities Authority, 2020), 10; Mithen, *After the Ice*, 76–81.

44. Lorenzo Nigro and Teresa Rinaldi, 'The Divine Spirit of Bees: A Note on Honey and the Origins of Yeast-Driven Fermentation', *Vicino Oriente*, 24 (2020), 185; Konstantinos D. Politis, 'Archaeology at the Lowest Place on Earth: Ghor es-Safi, Jordan', in Peilstöcker and Wolfram, *Life at the Dead Sea*, 263.

45. Hagay Hamer et al., 'The Neolithic Occupation of the Judean Desert Caves in Light of Recent Surveys and Excavations on the Eastern Cliffs', in *New Studies in the Archaeology of the Judean Desert* (Jerusalem: Israel Antiquities Authority, 2023), 150; Nir Hasson, 'Pisot megila gnuza vesheled ben 6,000 shana hitgalu be-midbar yehuda [Fragments of a Dead Sea Scroll and a 6,000-year-old skeleton in the Judean Desert]', *Haaretz*, 16 March 2021, https://www.haaretz.co.il/science/archeology/2021-03-16/ty-article-magazine/.premium/0000017f-f12f-df98-a5ff-f3af9a690000.

46. Ilkka Matero et al., 'The 8.2 ka Cooling Event Caused by Laurentide Ice Saddle Collapse', *Earth and Planetary Science Letters*, 473:1 (2017), 205–14; Frank H. Neumann and Wolfgang Zwickel, 'Settlements, Climate and Vegetation at the Dead Sea from the Neolithic until the Crusader Period', in Peilstöcker and Wolfram, *Life at the Dead Sea*, 78; Mithen, *After the Ice*, 87; Claudia Migowski et al., 'Holocene Climate Variability and Cultural Evolution in the Near East from the Dead Sea Sedimentary Record', *Quaternary Research*, 66 (2006), 425; Hamer et al., 'The Neolithic Occupation', 131–66.

47. Lorenzo Nigro, 'The Archaeology of Collapse and Resilience: Tell es-Sultan/Ancient Jericho as a Case Study', *ROSAPAT*, 11 (2014), 67.

48. Jutta Häser, 'Water Management in the Dead Sea Region', in Peilstöcker and Wolfram, *Life at the Dead Sea*, 95; Susanne Kerner, 'The Chalcolithic Period in the Dead Sea Area', in Peilstöcker and Wolfram, *Life at the Dead Sea*, 160; Kaptijn, *Life on the Watershed*.

49. Kerner, 'The Chalcolithic Period in the Dead Sea Area', 163; David Ilan and Yorke Rowan, 'Deconstructing and Recomposing the Narrative of Spiritual Life in the Chalcolithic of the Southern Levant (4500–3600 B.C.E.)', *Archaeological Papers of the American Anthropological Association*, 21:1 (2011), 90; Bernadette Drabsch, 'The Wall Art of Teleilat Ghassul: When, Where, Why, To Whom and By Whom?', *Expression*, 8 (2015), 50–57; 'Special Display: The Star of Ghassul', Rockefeller Archaeological

Museum, 7 October 2021, https://www.imj.org.il/en/exhibitions/special-display-star-ghassul.
50. Drabsch, 'The Wall Art of Teleilat Ghassul', 51.
51. Goren-Inbar et al., 'Evidence of Hominin Control', 726.
52. Dafna Langgut et al., 'The Origin and Spread of Olive Cultivation in the Mediterranean Basin: The Fossil Pollen Evidence', *The Holocene*, 29:5 (2019), 903.
53. Dafna Langgut and Yosef Garfinkel, '7000-Year-Old Evidence of Fruit Tree Cultivation in the Jordan Valley, Israel', *Scientific Reports*, 12:7463 (2022), 5–9.
54. Tamar Schick, *The Cave of the Warrior: A Fourth Millennium Burial in the Judean Desert* (Jerusalem: Israel Antiquities Authority, 1998).
55. Neumann and Zwickel, 'Settlements, Climate and Vegetation', 75.
56. Thomas Litt et al., 'Holocene Climate Variability in the Levant from the Dead Sea Pollen Record', *Quaternary Science Reviews*, 49 (2012), 100; Langgut and Garfinkel, '7000-Year-Old Evidence', 8.
57. Litt et al., 'Holocene Climate Variability', 103.
58. Gunnar Lehmann, 'Ancient Society and Economy at the Dead Sea from the Neolithic through the Persian Period', in Peilstöcker and Wolfram, *Life at the Dead Sea*, 195.
59. Migowski et al., 'Holocene Climate Variability', 425–6.
60. Eva Kaptijn, 'Surviving the Summer: Ancient Water Management in the Southern Jordan Valley as Compared to the Central Jordan Valley', in Peilstöcker and Wolfram, *Life at the Dead Sea*, 116. See also: Langgut and Garfinkel, '7000-Year-Old Evidence', 3.
61. Ian Kuijt et al., 'Pottery Neolithic Landscape Modification at Dhra'', *Antiquity*, 81 (2007), 107.
62. Häser, 'Water Management in the Dead Sea Region', 96; Mansour Shqiarat, 'History and Archaeology of Water Management in Jordan', *Scientific Culture*, 5:3 (2019), 44.
63. Dana Ackerfeld et al., 'Firing up the Furnace: New Insights on Metallurgical Practices in the Chalcolithic Southern Levant from a Recently Discovered Copper-Smelting Workshop at Horvat Beter (Israel)', *Journal of Archaeological Science*, 33 (2020), 1–14; Milena Gošić, 'Chalcolithic Metallurgy of the Southern Levant: Production Centers and Social Context', *Journal of the Serbian Archaeological Society*, 24 (2008), 69–70.
64. David Ussishkin, 'The Chalcolithic Temple in Ein Gedi: Fifty Years after its Discovery', *Near Eastern Archaeology*, 77:1 (2014), 15–26; Ilan and Rowan, 'Deconstructing and Recomposing', 93.
65. Kerner, 'The Chalcolithic Period in the Dead Sea Area', 167; Drabsch, 'The Wall Art of Teleilat Ghassul', 53.
66. Johanna Regev et al., 'Chronology of the Early Bronze Age in the Southern Levant: New Analysis for a High Chronology', *Radiocarbon*, 54:3/4 (2012), 525–66; Lorenzo Nigro, 'Archaeological Periodization vs Absolute Chronology', Proceedings of a workshop held in Palermo on 19 June 2017 (2019), 7.
67. Nigro, 'Jericho and the Dead Sea', 141; Lorenzo Nigro, 'Tell es-Sultan/Ancient Jericho in the Early Bronze Age II–III', *ROSAPAT*, 13 (2019), 82;

NOTES TO PP. 33-44

Elisabetta Gallo, 'Tell es-Sultan/Ancient Jericho in the Early Bronze Age II-III: A Population Estimate', *ROSAPAT*, 13 (2019), 109–12.
68. Nigro, 'Jericho and the Dead Sea', 141.
69. 'Bab edh-Dhra'', *Expedition to the Dead Sea Plain*, http://expeditiondeadseaplain.org/?page_id=27; Orendi, 'Development and Importance of Agrarian Resources', 127.
70. Lehmann, 'Ancient Society and Economy at the Dead Sea', 198; Häser, 'Water Management in the Dead Sea Region', 97; 'Numayra', *Expedition to the Dead Sea Plain*, http://expeditiondeadseaplain.org/?page_id=29; Steven Collins et al., 'The Tall al-Hammam excavation project: Season fifteen 2020 report', filed with the Department of Antiquities of the Hashemite Kingdom of Jordan, December 2021, 48.
71. Nigro, 'Tell es-Sultan/Ancient Jericho in the Early Bronze Age II–III', 85; Nigro, 'Archaeological Periodization', 19; Collins et al., 'The Tall al-Hammam Excavation Project', 7.
72. Nigro, 'Tell es-Sultan/Ancient Jericho in the Early Bronze Age II–III', 103; Nigro, 'Jericho and the Dead Sea', 145.
73. Nigro, 'Jericho and the Dead Sea', 145–8; Nigro, 'The Italian-Palestinian Expedition', 201; Lorenzo Nigro, 'Results of the Italian-Palestinian Expedition to Tell es-Sultan: At the Dawn of Urbanization in Palestine', *ROSAPAT*, 2 (2006), 28.
74. Häser, 'Water Management in the Dead Sea Region', 98; Orendi, 'Development and Importance of Agrarian Resources', 134; Lehmann, 'Ancient Society and Economy at the Dead Sea', 199; Politis, 'Archaeology at the Lowest Place on Earth', 264.
75. Elisa Joy Kagan et al., 'Dead Sea Levels during the Bronze and Iron Ages', *Radiocarbon*, 57:2 (2015), 247.

CHAPTER 2 A SITE OF GOD

1. The International River Jordan Water Company, *Souvenir Album of the Holy Land with Exceptionally Rare Views of the Sacred River Jordan* (New York, 1907).
2. Anson F. Rainey, 'Shasu or Habiru: Who Were the Early Israelites?', *Biblical Archaeology Review*, 34:6 (2008), 54. See also: Lehmann, 'Ancient Society and Economy at the Dead Sea', 199–200.
3. Eric E. Cline, *1177 B.C.: The Year Civilization Collapsed* (Princeton University Press, 2015), 94–5, 114–21.
4. Irina Russell and Magdel Le Roux, '"Global Warming" and the Movement of the Settlers to the Highlands of Palestine (circa 1200 – 1000 B.C.E.)', *Old Testament Essays*, 23:2 (2010), 324–5.
5. Josephus, *Jewish Antiquities*, books I–III (Cambridge, MA: Harvard University Press, 1998), 87.
6. Josephus, *The Jewish War*, books III–IV (Cambridge, MA: Harvard University Press: Harvard University Press, 1997), 299.
7. Joan Taylor, *The Essenes, the Scrolls, and the Dead Sea* (Oxford University Press, 2012), 231–2.
8. Roberto Tottoli, *Biblical Prophets in the Qur'ān and Muslim Literature* (London: Routledge, 2002), 28, 55; Guy Le Strange, *Palestine under the*

Moslems (London: Alexander P. Watt, 1890), 289–90; Zohar Amar, 'The Production of Salt and Sulphur from the Dead Sea Region in the Tenth Century according to at-Tamimi', *Palestine Exploration Quarterly*, 130 (1998), 4–5.

9. John of Würzburg, *Description of the Holy Land* (London, 1890), 60; Jacques de Vitry, *History of Jerusalem*, translated by Aubrey Stewart (London: Palestine Pilgrims' Text Society, 1896), 30.
10. Claude Conder, 'Report 18: Gilgal and the Plain of Jericho', 14 December 1874, PEF Archives, WS/CON/489.
11. Frederick Gardner Clapp, 'The Site of Sodom and Gomorrah. Diversity of Views', *American Journal of Archaeology*, 40:3 (1936), 344.
12. Konstantinos D. Politis (ed.), *Ancient Landscapes of Zoara I* (London: Routledge, 2020), 143–6.
13. Walter E. Rast and Thomas Schaub, 'Survey of the Southeast Plain of the Dead Sea', *Annual of the Department of Antiquities of Jordan*, 19 (1973), 19.
14. For more on excavations at the site, see: 'Bab adh-Dhra' ', *Expedition to the Dead Sea Plain*, http://expeditiondeadseaplain.org/?page_id=27.
15. Politis (ed.), *Ancient Landscapes*, 107.
16. The author's interview with Phillip Silvia.
17. Ted Bunch et al., 'A Tunguska Sized Airburst Destroyed Tall el-Hammam, a Middle Bronze Age City in the Jordan Valley near the Dead Sea', *Scientific Reports*, 11:1 (2021).
18. Ibid.
19. The author's correspondence with Lorenzo Nigro, February 2022.
20. Davide Bianchi, 'New Archaeological Discoveries in the Basilica of the Memorial of Moses, Mount Nebo', *Studies in the History and Archaeology of Jordan*, 13 (2019), 203–12.
21. Urbain Vermeulen and Jo van Steenbergen (eds), *Egypt and Syria in the Fatimid, Ayyubid and Mamluk Eras*, vol. III (Leuven: Peeters, 2001), 364; Hana Taragan, 'Holy Place in the Making: Maqam al-Nabi Musa in the Early Mamluk Period', *ARAM*, 18–19 (2006–2007), 621–39.
22. Edward Robinson, *Biblical Researches in Palestine, Mount Sinai and Arabia Petraea*, vol. II (Boston: Crocker & Brewster, 1841), 280.
23. Mark Twain, *The Innocents Abroad* (Hartford: American Publishing Company, 1869), 592, 594.
24. C.R. Conder and Horatio Herbert Kitchener, *The Survey of Western Palestine: Memoirs of the Topography, Orography, Hydrography and Archaeology*, vol. III (London, 1883), 223, 225.
25. Felicity Cobbing, 'John Garstang's Excavations at Jericho: A Cautionary Tale', *Strata*, 27 (2009), 63–77; Nigro, 'The Italian-Palestinian', 204. See also: Nigro, 'Tell es-Sultan/Ancient Jericho in the Early Bronze Age II–III', 103.
26. 'Ancient Jericho/Tell es-Sultan', UNESCO World Heritage Center, 20 October 2020, https://whc.unesco.org/en/tentativelists/6545/.
27. Russell and Le Roux, '"Global Warming" and the Movement of the Settlers', 332–3.
28. Israel Finkelstein, 'Jeroboam II in Transjordan', *Scandinavian Journal of the Old Testament*, 34:1 (2020), 22–3.

29. Davide Ventura, 'The Mesha Stele: A Reappraisal of a Forgery', preprint (2021), DOI: 10.6084/m9.figshare.14762265.v1.
30. Michael Press, 'The French Connection: Charles Clermont Ganneau and the Antiquities Trade Network in the Late Ottoman Levant', *Journal of Ancient Judaism*, 14 (2023), 198–200; Ruth Kark and Seth Frantzman, 'Empire, State and the Bedouin of the Middle East, Past and Present: A Comparative Study of Land and Settlement Policies', *Middle Eastern Studies*, 48:4 (2012), 490.
31. Chang-Ho Ji, 'A Moabite Sanctuary at Khirbat Ataruz, Jordan: Stratigraphy, Findings, and Archaeological Implications', *Levant*, 50:2 (2018), 173–210; Adam L. Bean and Christopher A. Rollston, 'Ataroth and the Inscribed Altar: Who Won the War Between Moab and Israel?', *The Torah*, 22 July 2020.
32. Joe Uziel and Ortal Chalafh, 'Archaeological Evidence of an Earthquake in the Capital of Judah', *City of David* (2021), 51–66.
33. Amotz Agnon, 'Pre-Instrumental Earthquakes along the Dead Sea Rift', in Z. Garfunkel et al. (eds), *Dead Sea Transform Fault System: Reviews* (Dordrecht: Springer, 2014), 207–61. See also: Alexander Fantalkin and Israel Finkelstein, 'The Sheshonq I Campaign and the 8th-Century BCE Earthquake – More on the Archaeology and History of the South in the Iron I-IIa', *Tel Aviv*, 33:1 (2006), 18–42; Aren M. Maeir and Joe Uziel (eds), *Tell es-Safi/Gath II: Excavations and Studies* (Münster: Zaphon, 2020), 35; Amanda Borschel-Dan, 'Archaeologists unearth 1st Jerusalem evidence of quake from Bible's Book of Amos', *Times of Israel*, 4 August 2021.
34. H.B. Tristram, *Bible Places; or, The Topography of the Holy Land* (London: Society for Promoting Christian Knowledge, 1884), 89.
35. Denys Pringle, *Pilgrimage to Jerusalem and the Holy Land, 1187–1291* (London: Routledge, 2012), 281; David Rapp and Hanan Isachar, *Churches and Monasteries in the Holy Land* (Herzliya: Apogee Press, 2014), 156–61.
36. Israel Rosenson, 'Beyin melach le-asphalt: Al shmotav shel yam ha-melach ba-mikrah [Between salt and asphalt: On the names of the Dead Sea in the Scriptures]', in Sigalit Rosemarin and Israel Rosenson (eds), *Da'at Lashon* (Jerusalem: Efrata, 2016), 236.
37. James Elmer Dean (ed.), *Epiphanius' Treatise on Weights and Measures: The Syriac Version* (University of Chicago Press, 1935), 34–5.
38. Cited in: Michael Philip Penn et al. (eds), *Invitation to Syriac Christianity: An Anthology* (University of California Press, 2022), 284.
39. Moses Shapira to Hermann Strack, 9 May 1883, cited in Ross Nichols's website: https://themosesscroll.com/moses-shapiras-9-may-1883-letter-to-hermann-strack/. See also: Ross Nichols, *The Moses Scroll* (Francisville, LA: Horeb Press, 2021).
40. Idan Dershowitz, 'The Valediction of Moses: New Evidence on the Shapira Deuteronomy Fragments', *Zeitschrift für die alttestamentliche Wissenschaft*, 133: 1 (2021), 1–22. See also: Shlomo Guil, 'The Shapira Scroll was an Authentic Dead Sea Scroll', *Palestine Exploration Quarterly*, 149:1 (2017), 6–27.
41. Claude R. Conder, 'To the Editor of The Times', *The Times*, 21 August 1883, 8.

NOTES TO PP. 62-71

42. Around the turn of the twenty-first century, a Hebrew inscription on a roughly polished limestone tablet, known as the 'Gabriel Inscription' or 'Gabriel Revelation', was sold by a Jordanian dealer to the collector David Jeselsohn. Analysis of the soil attached to the stone indicated that it had been found east of the Dead Sea, not far from the Lisan Peninsula. There is some debate regarding the interpretation of the text and its meaning, but it is usually dated to the first century BCE and is not attributed to the *yahad* community associated with the Dead Sea Scrolls. Tim McGirk, 'The Man Who Bought a Resurrection', *Time*, 8 July 2008; Torleif Elgvin, 'Eschatology and Messianism in the Gabriel Inscription', *Journal of the Jesus Movement in its Jewish Setting*, 1 (2014), 5–25.
43. Timothy H. Lim, *The Dead Sea Scrolls: A Very Short Introduction* (Oxford University Press, 2017), 19–20; Weston Fields, *The Dead Sea Scrolls: A Full History, 1947–1960* (Leiden: Brill, 2009).
44. The Damascus Document found in the caves of Qumran was already familiar to scholars, though in a slightly shorter version. It was one of the many documents that Solomon Schechter found in the Ben Ezra Synagogue, known as the Cairo Genizah, in 1897. Schechter was not sure whether this document was linked to the Samaritans, the Karaites or the Falashas. The Genizah version probably dates to the ninth to eleventh century CE. Stefan C. Reif, *Jews, Bible and Prayer* (Berlin: De Gruyter, 2017).
45. Géza Vermès, *Scripture and Tradition in Judaism: Haggadic Studies* (Leiden: Brill, 1983), 44–5.
46. Norman Golb, 'Khirbet Qumran and the Manuscripts of the Judaean Wilderness: Observations on the Logic of their Investigation', *Journal of Near Eastern Studies*, 49:2 (1990), 105.
47. Daniel Vainstub, 'The Covenant Renewal Ceremony as the Main Function of Qumran', *Religion*, 12:578 (2021), 1–26.
48. Sarit Anava et al., 'Illuminating Genetic Mysteries of the Dead Sea Scrolls', *Cell*, 181 (2020), 1–14.
49. Jodi Magness, *The Archaeology of the Holy Land* (Cambridge University Press, 2012), 122.
50. Daniel Burke, 'How forgers fooled the Bible Museum with fake Dead Sea Scroll fragments', *CNN*, 15 March 2020.

CHAPTER 3 A SITE OF WEALTH, WAR AND REFUGE

1. Jane Taylor, *Petra and the Lost Kingdom of the Nabataeans* (London: I.B. Tauris, 2001), 14–38.
2. Häser, 'Water Management in the Dead Sea Region', 98.
3. Diana Edelman, 'Different Sources, Different Views: Snapshots of Persian-Era Yehud Based on Texts and on Archaeological Data', *Estudios Bíblicos*, 76 (2018) 411–51.
4. Taylor, *Petra*, 37–8.
5. Diodorus Siculus, *Library of History*, vol. II (Cambridge, Cambridge, MA: Harvard University Press, 1935), 43–7; Strabo, *Geography* (Cambridge, Cambridge, MA: Harvard University Press, 1932), 295–6.

6. A. Nissenbaum, 'Dead Sea Asphalt in Ancient Egyptian Mummies – Why?', *Archaeometry*, 55:3 (2013), 563–8; J. Rullkötter and A. Nissenbaum, 'Dead Sea Asphalt in Egyptian Mummies: Molecular Evidence', *Naturwissenschaften*, 75 (1988), 618–21.
7. Jan Dušek, 'The Importance of the Wadi Daliyeh Manuscripts for the History of Samaria and the Samaritans', *Religions*, 11:63 (2020); Haggai Olshanetsky and Yael Escojido, 'Different from Others? Jews as Slave Owners and Traders in the Persian and Hellenistic Periods', *Sapiens Ubique Civis*, 1 (2020), 101–4.
8. Andrea Berlin, 'Zenon's Flour: Grains of Truth from Tel Kadesh', *Biblical Archaeology Review* (2019), 34–40; Olshanetsky and Escojido, 'Different from Others?', 104–9.
9. Stephen G. Rosenberg, 'Felicien de Saulcy and the Rediscovery of Tyros in Jordan', *Palestine Exploration Quarterly*, 138:1 (2006), 35–41; Magness, *The Archaeology*, 74.
10. Ze'ev Meshel, 'The Nabataean "Rock" and the Judaean Desert Fortresses', *Israel Exploration Journal*, 50:1/2 (2000), 109–15.
11. Ministry of Foreign Affairs, 'First evidence of the Maccabean Revolt against the Greek Seleucid Kingdom found in the Judean Desert', 13 December 2022, https://www.gov.il/en/departments/general/evidence-found-of-the-maccabean-revolt-against-greek-seleucid-kingdom-13-dec-2022; Eitan Klein et al., 'A Hoard of Ptolemaic Coins Found in Murabba'at Cave II in the Judean Desert', *Atiqot*, 112 (2023), 53–92.
12. Yizhar Hirschfeld and Donald T. Ariel, 'A Coin Assemblage from the Reign of Alexander Jannaeus Found on the Shore of the Dead Sea', *Israel Exploration Journal*, 55:1 (2005), 66–89.
13. Günter Garbrecht and Yehuda Peleg, 'The Water Supply of the Desert Fortresses in the Jordan Valley', *The Biblical Archaeologist*, 57:3 (1994), 161–70.
14. Magness, *The Archaeology*, 204; Meshel, 'The Nabataean "Rock"', 110.
15. Kathryn Louise Gleason, 'A Garden Excavation in the Oasis Palace of Herod the Great at Jericho', *Landscape Journal*, 12:2 (1993), 156–67; Magness, *The Archaeology*, 185. For more on Herod's construction projects, see: Ehud Netzer, *The Architecture of Herod, the Great Builder* (Tübingen: Mohr Siebeck, 2020).
16. Josephus, *History of the Jewish War against the Romans*, book 4, chapter 8; Garbrecht and Peleg, 'The Water Supply', 165–9.
17. Győző Vörös, 'Machaerus: The Golgotha of John the Baptist', *Revue Biblique*, 119:2 (2012), 232–70.
18. Josephus, *History of the Jewish War against the Romans*, book 7, 4; Magness, *The Archaeology*, 214.
19. Ulrich Jasper Seetzen, *Reisen durch Syrien, Palästina, Phönicien, die Transjordan-länder, Arabia Petraea und Unter-Aegypten* (Berlin, 1859), 371–2.
20. Jonathan Roth, 'The Length of the Siege of Masada', *Scripta Classica Israelica*, 14 (1995), 87–110; Hai Ashkenazi et al., 'The Roman Siege System of Masada: A 3D Computerized Analysis of a Conflict Landscape', *Journal of Roman Archaeology* (2024), 1–26.
21. Shemaryahu Talmon, 'Hebrew Written Fragments from Masada', *Dead Sea Discoveries*, 3:2 (1996), 168–77.

NOTES TO PP. 83-93

22. Anava et al., 'Illuminating Genetic Mysteries', 1227.
23. Jodi Magness, *Masada: From Jewish Revolt to Modern Myth* (Princeton University Press, 2019).
24. Stéphanie E. Binder, Michael Lazar and Emmanuel Nantet, 'Measurements and Shape of the Dead Sea in the Hellenistic and Roman Periods', *Israel Exploration Journal*, 69:2 (2019), 176.
25. Aristotle, *Meteorologica*, vol. II, 159.
26. Pliny the Elder, *Natural History*, book V (Cambridge, Cambridge, MA: Harvard University Press, 1938), 274.
27. John Wilson, *The Lands of the Bible: Visited and Described in an Extensive Journey*, vol. II (Edinburgh, 1847), 8–10. See also: Jacob E. Spafford, 'Around the Dead Sea by Motor Boat', *The Geographical Journal*, 39:1 (1912), 39.
28. Gideon Hadas, 'The Balsam "Afarsemon" and Ein Gedi during the Roman-Byzantine Period', *Revue Biblique* (2007), 161–73; Eleni Manolaraki, 'Hebraei Liquores: The Balsam of Judaea in Pliny's Natural History', *The American Journal of Philology*, 136:4 (2015), 633–67.
29. Hadas, 'The Balsam'; Manolaraki, 'Hebraei Liquores'; Hannah Cotton, 'The Date of the Fall of Masada: The Evidence of the Masada Papyri', *Zeitschrift für Papyrologie und Epigraphik*, 78 (1989), 157–62; Taylor, *The Essenes*, 225.
30. Cotton, 'The Date'; Hadas, 'The Balsam'; Yizhar Hirschfeld, 'Ein Gedi: A Large Jewish Village', *Qadmoniot*, 128 (2005), 62–87.
31. Sarah Sallon et al., 'Germination, Genetics, and Growth of an Ancient Date Seed', *Science*, 320 (2008), 1464; Sarah Sallon et al., 'Origins and Insights into the Historic Judean Date Palm based on Genetic Analysis of Germinated Ancient Seeds and Morphometric Studies', *Science Advances*, 6 (2020).
32. 'Six new ancient date trees', Arava Institute website, https://arava.org/arava-research-centers/arava-center-for-sustainable-agriculture/methuselah/6-new-ancient-date-trees/.
33. Correspondence with the author, 1 December 2023.
34. Father Roland de Vaux, *The Dead Sea Scrolls, A Full History* (Leiden: Brill, 2009), 124.
35. G. Graystone, 'The Dead Sea Scrolls – II. Wadi Murabba'at,' *Scripture*, 7:3 (1955), 66–76.
36. 'The Newest of the Dead Sea Scrolls', *Time*, 24 January 1977, 57–8; Neil Asher Silberman, *A Prophet from Amongst You: The Life of Yigael Yadin* (Boston: Addison-Wesley, 1993), 305–6.
37. Yigael Yadin, *Bar Kokhba: The Rediscovery of the Legendary Hero of the Last Jewish Revolt against Imperial Rome* (London: Weidenfeld and Nicolson, 1971).
38. Cassius Dio, *Roman History*, vol. VIII: books *61–70* (Cambridge, Cambridge, MA: Harvard University Press, 1925), 451.
39. Hanan Eshel and Boaz Zissu, *The Bar Kokhba Revolt: The Archaeological Evidence* (Jerusalem: Yad Ben Zvi, 2019), 14.

NOTES TO PP. 93-101

40. Jonathan Bourgel, 'Ezekiel 40–48 as a Model for Bar Kokhba's Title "Nasi Israel"?', *Journal of Ancient Judaism*, 14 (2023), 446–81; Hannah Cotton, 'Ein Gedi between the Two Revolts', *Scripta Classica Israelica*, 20 (2001), 139–54.
41. Yadin, *Bar Kokhba*, 133.
42. Ibid., 222–53.
43. David Gritten, 'Dead Sea reveals four 1,900-year-old Roman swords in cave', *BBC*, 6 September 2023, https://www.bbc.co.uk/news/world-middle-east-66728207.
44. Hagith Sivan, *Palestine in Late Antiquity* (Oxford University Press, 2008), 40.
45. Magness, *The Archaeology*, 342; Sivan, *Palestine in Late Antiquity*, 58; Taylor, *The Essenes*, 263, 300.
46. Eli Ashkenazi et al., 'Poplar Trees in Israel's Desert Regions: Relicts of Roman and Byzantine Settlement', *Journal of Arid Environments*, 193 (2021).
47. Philip Mayerson, 'The Saracens and the *Limes*', *Bulletin of the American Schools of Oriental Research*, 262 (1986), 35–47; Thomas Parker, 'An Empire's New Holy Land: The Byzantine Period', *Near Eastern Archaeology*, 62:3 (1999), 157; Hans-Peter Kuhnen, 'Guarding the Dead Sea: Military Concepts and Sites between Herod and Justinian', in Peilstöcker and Wolfram, *Life at the Dead Sea*, 230.
48. Rehav Rubin, 'The Debate Over Climatic Changes in the Negev, Fourth-Seventh Centuries C.E.', *Palestine Exploration Quarterly*, 121:1 (1989), 76–7.
49. R. Bookman et al., 'Late Holocene Lake Levels of the Dead Sea', *Geological Society of America Bulletin*, 116 (2004), 555–71; Frank H. Neumann et al., 'Vegetation History and Climate Fluctuations on a Transect along the Dead Sea West Shore and their Impact on Past Societies over the Last 3500 Years', *Journal of Arid Environments*, 74:7 (2010), 758–9; Adam Izdebski, 'Why Did Agriculture Flourish in the Late Antique East?' *Millennium*, 8:1 (2011), 308–9; Neumann and Zwickel, 'Settlements, Climate and Vegetation', 78; Nurit Weber et al., 'Gypsum Deltas at the Holocene Dead Sea Linked to Grand Solar Minima', *Geophysical Research Letters* (2021), 7.
50. Yizhar Hirschfeld, 'A Climatic Change in the Early Byzantine Period? Some Archaeological Evidence', *Palestine Exploration Quarterly*, 136:2 (2004), 133–49.
51. Beatrice Leal, 'A Reconsideration of the Madaba Map', *Gesta*, 57:2 (2018), 123–43.
52. Nigel F. Hepper and Joan E. Taylor, 'Date Palms and Opobalsam in the Madaba Mosaic Map', *Palestine Exploration Quarterly*, 136:1 (2004), 35–44; Zaraza Friedman, 'Sailing in the Dead Sea: Madaba Map Mosaic', in L. Daniel Chrupcala (ed.), *Christ Is Here! Studies in Biblical and Christian Archaeology in Memory of Michele Piccirillo* (Milano: Edizioni Terra Santa, 2012), 381–94.
53. Asaf Oron et al., 'Two Artificial Anchorages off the Northern Shore of the Dead Sea', *International Journal of Nautical Archaeology*, 44:1 (2015), 89.

NOTES TO PP. 101-109

54. Friedman, 'Sailing in the Dead Sea', 387; Gideon Hadas, Nili Liphschitz and Georges Bonani, 'Two Ancient Wooden Anchors from Ein Gedi, on the Dead Sea, Israel', *International Journal of Nautical Archaeology*, 34:2 (2005), 299-307; Jean-Baptiste Humbert, 'Qumran and Machaerus on a Hasmonean Axis', in Peilstöcker and Wolfram, *Life at the Dead Sea*, 327.
55. Sivan, *Palestine in Late Antiquity*, 58–62.
56. Hirschfeld, 'Ein Gedi', 143–5.
57. Juan Manuel Tebes, 'Beyond Petra: Nabataean Cultic and Mortuary Practices and the Cultural Heritage of the Negev and Edom', *Jordan Journal for History and Archaeology*, 14:4 (2020), 333–47; Konstantinos D. Politis, 'The Discovery and Excavation of the Khirbet Qazone Cemetery and its Significance Relative to Qumran', in Jean-Baptiste Humbert and Jürgen Zangenberg (eds), *Qumran: The Site of the Dead Sea Scrolls: Archaeological Interpretations and Debates* (Leiden: Brill, 2006), 213–19; Sivan, *Palestine in Late Antiquity*, 243–6.
58. Sivan, *Palestine in Late Antiquity*, 36, 57–60.
59. Kuhnen, 'Guarding the Dead Sea', 230.
60. Hirschfeld, 'Ein Gedi', 144–5; Itamar Taxel, 'The Byzantine-Early Islamic Transition in the Dead Sea Region', in Peilstöcker and Wolfram, *Life at the Dead Sea*, 306.

CHAPTER 4 A SITE OF DECLINE, NEW BEGINNINGS AND MYTHS

1. Cited in: Le Strange, *Palestine under the Moslems*, 289–90.
2. De Vitry, *The History of Jerusalem*, 30.
3. Michael J. Decker, *The Byzantine Dark Ages* (Bloomsbury, 2016), 11.
4. Michael McCormick, *Origins of the European Economy: Communications and Commerce, AD 300–900* (Cambridge University Press, 2001), 40; Gideon Avni, *The Byzantine–Islamic Transition in Palestine* (Oxford University Press, 2014), 328–9.
5. Weber et al., 'Gypsum Deltas', 7–8.
6. Decker, *The Byzantine Dark Ages*, 149; Izdebski, 'Why Did Agriculture Flourish', 291–312; Dominik Fleitmann et al., 'Droughts and Societal Change: The Environmental Context for the Emergence of Islam in Late Antique Arabia', *Science*, 376:6599 (2022), 1317–21.
7. Gideon Lev, 'Etz zayit atik movil hokrim le-shorshai ha-haklaot ba-negev' [An ancient olive tree leads scholars to the roots of agriculture in the Negev]', *Haaretz*, 14 April 2023; Rachel Blevis et al., 'Fish in the Desert: Identifying Fish Trade Routes and the Role of Red Sea Parrotfish (Scaridae) during the Byzantine and Early Islamic Periods', *Journal of Archaeological Science: Reports*, 36 (2021); J. Brett Hill, *Human Ecology in the Wadi Al-Hasa: Land Use and Abandonment through the Holocene* (University of Arizona Press, 2006), 53–4.
8. Taxel, 'The Byzantine-Early Islamic Transition', 299.
9. David Neev and Kenneth Orris Emery, 'The Dead Sea: Depositional Process and Environments of Evaporates', *Geological Survey of Israel*, bulletin no. 41 (1967), 30.

10. Diana K. Davis, 'Restoring Roman Nature: French Identity and North African Environmental History', in D.K. Davis and E. Burke (eds), *Environmental Imaginaries of the Middle East and North Africa* (Ohio University Press, 2011), 60–86.
11. Henri Pirenne, *Mohammed and Charlemagne* (London, 1939).
12. Marcus Milwright, 'Balsam in the Mediaeval Mediterranean: A Case Study of Information and Commodity Exchange', *Journal of Mediterranean Archaeology*, 14:1 (2001), 3–23; Zohar Amar and Efraim Lev, 'Trends in the Use of Perfumes and Incense in the Near East after the Muslim Conquests', *Journal of the Royal Asiatic Society*, 23:1 (2013), 27.
13. Avni, *The Byzantine–Islamic Transition in Palestine*, 202; Taxel, 'The Byzantine-Early Islamic Transition', 297–300.
14. Taxel, 'The Byzantine-Early Islamic Transition', 310.
15. D.H.K. Amiran et al., 'Earthquakes in Israel and Adjacent Areas: Macroseismic Observations since 100 BCE', *Israel Exploration Journal*, 44 (1994), 260–305; Mohamed Reda Sbeinati et al., 'The Historical Earthquakes of Syria: An Analysis of Large and Moderate Earthquakes from 1365 B.C. to 1900 A.D.', *Annals of Geophysics*, 48:3 (2005), 362–3.
16. Donald Whitcomb, 'The Jericho Mafjar Project: Palestine-University of Chicago research at Khirbet el-Mafjar', in Sparks et al. (eds), *Digging Up Jericho* (2020), 247–54.
17. Ronnie Ellenblum, *Frankish Rural Settlements in the Latin Kingdom of Jerusalem* (Cambridge University Press, 1998), 20–22; Avni, *The Byzantine–Islamic Transition in Palestine*, 332.
18. James N. Kelso, 'Excavations at Khirbet en-Nitla near Jericho', *Bulletin of the American School of Oriental Research*, 121 (1951), 6; Avni, *The Byzantine–Islamic Transition in Palestine*, 228; Itamar Taxel, 'Pots and People from East to West: A Rare Early Islamic Glazed Lidded Bowl from a Monastic Site near Jericho, and its Cultural and Historical Context', *Archäologischer Anzeiger* (2014), 85–99.
19. Menahem Kister and Meir Jacob Kister, 'Al yehodai arav – he'arot [On the Jews of Arabia – Comments]', *Tarbiz* (1979), 235; Hanan Eshel, 'Teiaruch beit ha-knesset be-yericho [Dating the Synagogue in Jericho]', *Qardom*, 5 (1983), 178–80; Taxel, 'The Byzantine-Early Islamic Transition', 308.
20. Bianchi, 'New Archaeological Discoveries', 210–11.
21. K.D. Politis, 'The Monastery of Aghios Lot at Deir 'Ain 'Abata in Jordan', in *Byzanz: das Römerreich im Mittelalter* (Mainz: Schnell & Steiner, 2010), 1–24.
22. Amar, 'The Production', 3–5.
23. Jaime Wisniak, 'Dyes from Antiquity to Synthesis', *Indian Journal of History and Science*, 39:1 (2004), 82–4; Le Strange, *Palestine under the Moslems*, 18, 31, 289; Konstantinos D. Politis, 'The Sugar Industry in the Ghawr as-Safi, Jordan', *Studies in the History and Archaeology of Jordan*, 11 (2013), 470; Donald Whitcomb, 'Jericho Mafjar Project', 2011–2012 Report for the Oriental Institute, 64; Taxel, 'The Byzantine-Early Islamic Transition', 296.
24. Tsugitaka Sato, *Sugar in the Social Life of Medieval Islam* (Leiden: Brill, 2015), 19–23; Judith Bronstein, Edna J. Stern & Elisabeth Yehuda, 'Franks,

Locals and Sugar Cane: A Case Study of Cultural Interaction in the Latin Kingdom of Jerusalem', *Journal of Medieval History*, 45:3 (2019), 316–30.
25. De Vitry, *The History of Jerusalem*, 30.
26. Hamdan Taha, 'The Sugarcane Industry in Jericho, Jordan Valley', in Konstantinos D. Politis (Ed.), *The Origins of the Sugar Industry and the Transmission of Ancient Greek and Medieval Arab Science and Technology from the Near East to Europe* (Athens: National and Kapodistriako University of Athens, 2015), 51–77; Politis, 'The Sugar Industry', 467–80; Philip Slavin, '"With a Grain of Sugar": Native Agriculture and Colonial Capitalism in the Frankish Levant, c. 1100–1300', *Crusades*, 22:1 (2023), 29; Richard Jones, *Sweet Waste: Medieval Sugar Production in the Mediterranean Viewed from the 2002 Excavation at Tawahin es-Sukkar, Safi, Jordan* (Glasgow: Potingair Press, 2017); Felix Fabri, *Evagatorium in Terrae Sanctae, Arabiae et Aegypti Peregrinationem*, vol. II (Stuttgart, 1843), 50; Eliyahu Ashtor, *Technology, Industry and Trade: The Levant versus Europe, 1250–1500* (Ashgate: Routledge, 1992), 228–62.
27. Edward E. Curtis, 'The Ghawarna of Jordan: Race and Religion in the Jordan Valley', *Journal of Islamic Law and Culture*, 13:2 (2011), 195.
28. Politis, 'Archaeology at the Lowest Place', 278; Yaacov Lev, 'David Ayalon (1914–1998) and the History of Black Military Slavery in Medieval Islam', *Der Islam*, 90:1 (2013), 21–43.
29. Le Strange, *Palestine under the Moslems*, 290.
30. Fulcher of Chartres, *A History of the Expedition to Jerusalem 1095–1127* (University of Tennessee Press, 1969), 146.
31. William of Malmesbury, *Gesta Regum Anglorum, The History of the English Kings* (Oxford: Clarendon Press, 1998), 673.
32. Jochen G. Schenk, 'Nomadic Violence in the First Latin Kingdom of Jerusalem and the Military Orders', *Reading Medieval Studies*, 36 (2010), 41.
33. Le Strange, *Palestine under the Moslems*, 65.
34. 'Mauricius Montis Regalis possessor et dominus', 1152, glossary available on the Revised Regesta Regni Hierosolymitani database, http://crusades-regesta.com/database?search_api_views_fulltext=Mauricius+&field_institution_recipient=&field_grantor=&field_recepient=&field_year_1=&field_year=&field_term_type_field_term_title=.
35. Asaf Oron et al., 'A New Type of Composite Anchor Dated to the Fatimid-Crusader Period from the Dead Sea, Israel', *International Journal of Nautical Archaeology*, 37:2 (2008), 295–301.
36. Ellenblum, *Frankish Rural Settlements*, 191–2; Cristina Violante, 'The Tempo of Water', *Journal of Palestine Studies*, 51:4 (2022), 70–1.
37. I. Istakhri, *al-Masalik wa al-Mamalik* [Roads and Kingdoms] (Beirut, 2004), 7–28.
38. Amar, 'The Production', 7; Ronnie Ellenblum, *The Collapse of the Eastern Mediterranean: Climate Change and the Decline of the East, 950–1072* (Cambridge University Press, 2012).
39. Fulcher of Chartres, *A History of the Expedition to Jerusalem*, 145–6.
40. De Vitry, *The History of Jerusalem*, 30. See also: Denys Pringle, 'Templar Castles on the Road to the Jordan', in Malcolm Barber (ed.), *The Military*

Orders: Fighting for the Faith and Caring for the Sick (Ashgate: Routledge, 1994), 148–66; William of Tyre, *A History of the Deeds Done beyond the Sea*, vol. I (New York, 1943), 339–40.
41. Pringle, 'Templar Castles', 148–66.
42. Alex Mallett, 'A Trip Down the Red Sea with Reynald of Châtillon', *Journal of the Royal Asiatic Society*, 18:2 (2008), 141.
43. Marcus Milwright, *The Fortress of the Raven: Karak in the Middle Islamic Period (1100–1650)* (Leiden: Brill, 2008), 27–37.
44. Nicholas Morton, *The Mongol Storm* (London: Basic Books, 2023), 214–32; Bethany J. Walker, 'Mamluk Investment in Southern Bilād Al-Shām in the Eighth/Fourteenth Century: The Case of Ḥisbān', *Journal of Near Eastern Studies*, 62:4 (2003), 243.
45. Taragan, 'Holy Place in the Making', 627.
46. Hirschfeld, 'A Climatic Change', 144; Weber et al., 'Gypsum Deltas', 7; Graystone, 'The Dead Sea Scrolls', 66.
47. Ashtor, *Technology, Industry and Trade*, 275–6, 284; Walker, 'Mamluk Investment', 248–9.
48. Wolf-Dieter Hütteroth and Kamal Abdulfattah, *Historical Geography of Palestine, Transjordan and Southern Syria in the Late 16th Century* (Erlangen, 1977), 54–9.
49. Joan Taylor, 'The Dead Sea in Western Travellers' Accounts from the Byzantine to the Modern Period', *Strata*, 27 (2009), 15.
50. Nathan Marcus Adler, *The Itinerary of Benjamin of Tudela: Critical Text, Translation and Commentary* (London, 1907), 23; Burchard of Mount Sion, A.D. *1280*, trans. by Aubrey Stewart (London, 1896), 59.
51. Fabri, *Evagatorium in Terrae Sanctae*, 144–6; Taylor, 'The Dead Sea', 17; Arie Nissenbaum, 'The Dead Sea Monster', *International Journal of Salt Lake Research*, 1:2 (1992), 4–5.
52. Nissenbaum, 'The Dead Sea Monster', 5.
53. Taylor, 'The Dead Sea', 26; Nissenbaum, 'The Dead Sea Monster', 6; Jacques de Vitry, *Historia Orientalis* (Duaci, 1596), 187 [scanned by Internet Archive].
54. Rehav Rubin, 'Greek-Orthodox Maps of Jerusalem from the Seventeenth and Eighteenth Centuries', *e-Perimetron*, 8:3 (2013), 106–32; Pnina Arad, 'Landscape and Iconicity: Proskynetaria of the Holy Land from the Ottoman Period', *The Art Bulletin*, 100:4 (2018), 62–80.
55. *Proskynētaria tōn Hagiōn Topōn: Deka hellēnika cheirographa 16ou–18ou ai* [Pilgrimages of the Holy Places: Ten Greek Manuscripts 16th–18th Centuries] (Thessaloniki, 1986), 104.
56. *Proskynētaria*, 104, 119, 163, 181, 214. Martin Kabátník, who travelled from Bohemia to Syria, Palestine and Egypt in 1491–1492, shared some of the impressions of the Greek Orthodox *Proskynetaria*.
57. Kimberley Thomas, 'Beyond the Plantation: Salt, Turks Islands, Bermuda and the British Atlantic World, 1660s–1850s' (PhD thesis, University of Warwick, 2019), 64–7.
58. Thomas Fuller, *A Pisgah Sight of Palestine and the Confines Thereof* (London: John Williams, 1650).

NOTES TO PP. 128–137

59. Taylor, 'The Dead Sea', 15, 24.
60. Thomas Wright (ed.), *Early Travels in Palestine* (London, 1848), 453–4.
61. Fredrik Hasselquist, *Voyages and Travels in the Levant in the Years 1749, 50, 51, 52* (London, 1766), 284.
62. Richard Pococke, *A Description of the East and Some Other Countries* (London, 1745), 36–8.

CHAPTER 5 A SITE OF SCIENCE, EXPLORATION AND DIVERGING NORMS

1. E.W.G. Masterman, 'Summary of the Observations on the Rise and Fall of the Level of the Dead Sea, 1900–1913', *Palestine Exploration Quarterly*, 45:4 (1913), 194. See also: Haim Goren, *Dead Sea Level: Science, Exploration and Imperial Interests in the Middle East* (London: I.B. Tauris, 2011), 148.
2. De Vitry, *The History of Jerusalem*, 29.
3. Poulain de Bossay, 'Note sur quelques explorations à faire en Syrie, en Palestine, et dans l'Arabie-Pétrée', *Bulletin de la Société de Géographie*, 29 (1838), 47; Jules de Bertou, 'Notes on a Journey from Jerusalem by Hebron, the Dead Sea, El Ghór, and Wádí Arabah to Akabah, and Back by Petra; in April, 1838', *The Journal of the Royal Geographical Society of London*, 9 (1839), 286.
4. De Bertou, 'Notes on a Journey', 286.
5. Goren, *Dead Sea Level*, 205–6; 'On the Dead Sea and Some Positions in Syria', *The Journal of the Royal Geographical Society of London*, 7 (1837), 456.
6. Goren, *Dead Sea Level*, 213–14, 258.
7. Col. Sir Henry James, 'Report on the Levelling from the Mediterranean to the Dead Sea', 3 May 1866, PEF Archives, JER/WIL/20 1866.
8. Goren, *Dead Sea Level*, 271.
9. Conder and Kitchener, *The Survey of Western Palestine*, 221.
10. William Francis Lynch, *Narrative of the United States' Expedition to the River Jordan and the Dead Sea* (Philadelphia, 1850), 270; Carmelo Pappalardo, 'Una Lettera di Paolo Bajnotti a Cristoforo Negri nel contesto dell'esplorazione del Mar Morto nel XIX secolo', *Bollettino della Società Geografica Italiana*, 14:4 (2021), 54; Cippora Klein, 'Shinuyei hamiflas shel yam-ha-melach [Changes in the Level of the Dead Sea]', in Mordechai Naor (ed.), *Yam ha-Melach ve-Midbar Yehuda, 1900–1967* (Jerusalem, 1990), 39–40; John Gray Hill, 'The Dead Sea', *Palestine Exploration Fund Quarterly Statement* (1900), 273–82; Oron et al., 'Two Artificial Anchorages', 86.
11. Charles F. Tyrwhitt Drake, 'Mr Tyrwhitt Drake's Reports', 21 March 1874, *Palestine Exploration Fund Quarterly Statement* (1875), 187–90; Hill, 'The Dead Sea', 274–5; Masterman, 'Summary of the Observations', 196.
12. Charles W. Wilson (ed.), *Picturesque Palestine, Sinai and Egypt*, vol. III (London, 1880), 207.
13. Masterman, 'Summary of the Observations', 195.
14. Wright (ed.), *Early Travels in Palestine*, 453.

257

NOTES TO PP. 138-147

15. Hasselquist, *Voyages and Travels*, 127–9, 284.
16. Goren, *Dead Sea Level*, 140.
17. Ibid., 131.
18. Henry Baker Tristram, *The Land of Israel: A Journal of Travels in Palestine* (London, 1865), v.
19. Ibid., 201–2.
20. Ibid., 227.
21. Ibid., 245.
22. Goren, *Dead Sea Level*, 145, 154.
23. Louis Lortet, *La Syrie d'aujourd'hui: voyages dans la Phénicie, le Liban, et la Judée, 1875–1880* (Paris, 1884), 432; Louis Lortet, 'Researches on the Pathogenic Microbes of the Mud of the Dead Sea', *Palestine Exploration Quarterly*, 24:1 (1892), 48–50.
24. Benjamin Wilkansky, 'Life in the Dead Sea', *Nature*, 12 September 1936; Benjamin Elazari-Volcani, 'Algae in the Bed of the Dead Sea', *Nature*, 22 June 1940; Benjamin Elazari Volcani, 'Bacteria in the Bottom Sediments of the Dead Sea', *Nature*, 4 September 1943; Ahron Oren and Antonio Ventosa, 'Benjamin Elazari Volcani (1915–1999): Sixty–three Years of Studies of the Microbiology of the Dead Sea', *International Microbiology*, 2 (1999), 195–8.
25. Naomi Oreskes, 'Earth Science: How Plate Tectonics Clicked', *Nature*, 501 (2013), 27–9.
26. William Allen, *The Dead Sea: A New Route to India*, vol. I (London: Longman, 1855), 243.
27. Tristram, *Land of Israel*, 197.
28. Ibid., 218.
29. Wilfrid H. Hudleston, *The Geology of Palestine* (London, 1885), 66–7, 70–71.
30. Edward Hull, *The Survey of Western Palestine: Memoir on the Physical Geology and Geography of Arabia Petraea, Palestine, and Adjoining Districts* (London, 1886), 103.
31. Ibid., 104. Hull gives credit to observations made in: Tristram, *Land of Israel*, 326.
32. Hull, *Survey of Western Palestine*, 120.
33. E.W.G. Masterman, 'The Physical History of the Dead Sea,' *The Biblical World*, 25:4 (1905), 251.
34. Huntington, *Palestine*, 180, 188.
35. Cited in: Enrico Bonatti, Anna Cipriani and Luca Lupi, 'The Red Sea: Birth of an Ocean', in Najeeb M.A. Rasul and Ian C.F. Stewart (eds), *The Red Sea* (Heidelberg: Springer, 2015), 31.
36. Post by Tomas Pueyo, 9 May 2023, https://twitter.com/tomaspueyo/status/1655941795971469312.
37. 'Canals: Plan for a Ship Canal to India by the Dead Sea', 15 February 1853, The National Archives of the UK [hereafter: TNA], HO 45/4946; Goren, *Dead Sea Level*, 120.
38. Allen, *The Dead Sea*, vol. I, vi, 359.
39. Hudleston, *Geology of Palestine*, 74.

40. Samuel E. Willner, 'Maritime Navigation, Energy Security, and Peace Dividend: Exploring the Historic Developments of the Dead Sea Conveyance Project', *Journal of the Middle East and Africa*, 8:1 (2017), 117.
41. Theodor Herzl, *Old New Land*, translated by Lotta Levensohn (Princeton: Markus Wiener, 2007), 242.
42. Walter Scott, *The Talisman* (Oxford University Press, 1912), 1–3.
43. *The Scapegoat*, National Museums Liverpool, https://www.liverpoolmuseums.org.uk/artifact/scapegoat.
44. Clinton Bailey, 'Dating the Arrival of Bedouin Tribes in Sinai and the Negev', *Journal of the Economic and Social History of the Orient*, 28:1 (1985), 22–4.
45. Charles F. Tyrwhitt Drake, 'Tyrwhitt Drake's Report', May 1874, *Palestine Quarterly Statement* (1875), 27–8.
46. John Wilson, *The Lands of the Bible Visited and Described*, vol. II (Edinburgh, 1847), 6.
47. Yoav Alon, *The Making of Jordan: Tribes, Colonialism and the Modern State* (London: I.B. Tauris, 2009), 162; Curtis, 'The Ghawarna of Jordan', 193–209.
48. Lynch, *Narrative of the United States' Expedition*, 79.
49. Allen, *The Dead Sea*, 225–7.
50. Tristram, *Land of Israel*, 170, 206.
51. Ibid., 192, 223.
52. Ibid., 337.
53. Twain, *The Innocents Abroad*, 587–8.
54. Drake, 'Tyrwhitt Drake's Report', May 1874, 27.
55. Laurence Oliphant, *Land of Gilead* (Edinburgh and London, 1880), 305, 310–11.
56. Huntington, *Palestine*, 190–91.
57. Tristram, *Land of Israel*, 192.
58. Archibald Forder, *With the Arabs in Tent and Town* (London: Marshal Brothers, 1902), 192.
59. Tristram, *Land of Israel*, 337.
60. Hasselquist, *Voyages and Travels*, 129.
61. John Carne, *Syria, the Holy Land & Asia Minor Illustrated*, vol. 3 (London: Fisher, 1840), 69.
62. Robinson, *Biblical Researches in Palestine*, vol. II, 279–81.
63. Wilson, *Lands of the Bible*, vol. II, 7.
64. Tristram, *Land of Israel*, 207.
65. Eugene Rogan, *Frontiers of the State in the Late Ottoman Empire: Transjordan, 1850–1921* (Cambridge University Press, 1999), 62–3; Forder, *With the Arabs*, 193–5; Jacob Norris, 'Toxic Waters: Ibrahim Hazboun and the Struggle for a Dead Sea Concession, 1913–1948', *Jerusalem Quarterly* (2011), 27; Hila Tal, *Mahapecha Be-Sdom* [Renaissance at Sodom] (Ben Gurion Research Institute, 2010), 15–17.
66. Masterman, 'Physical History', 252.
67. Elhanan Leib Lewinsky, *Masa le-Eretz Yisrael be-shnat TT la-elef ha-shishi* [Voyage to the Land of Israel in the Year 5800 (2040)] (Odessa, 1892), 44.

NOTES TO PP. 159-169

68. Herzl, *Old New Land*, 168.
69. Norris, 'Toxic Waters', 28.
70. Moshe Novomeysky, *Given to Salt: The Struggle for the Dead Sea Concession* (London, 1958), 11.

CHAPTER 6 A SITE OF EXTRACTION, CONFLICT AND WATER SCARCITY

1. Charles Clermont-Ganneau, 'New Hebrew mosaic unearthed by Turkish shell: Palestine campaign echo', *The Times*, 10 October 1919, 9.
2. Rémy Courcier, Jean-Philippe Venot and François Molle, 'Lower Jordan River Basin (in Jordan): Changes in Water Use and Projections (1950–2025)', International Water Management Institute, report no. 9 (2005), 2; Munther J. Haddadin, 'A Jordanian Socio-Legal Perspective on Water Management in the Jordan River–Dead Sea Basin', in Clive Lipchin, Deborah Sandler and Emily Cushman (eds), *The Jordan River and Dead Sea Basin: Cooperation amid Conflict* (Dordrecht: Springer, 2009), 42.
3. Standard Oil Co. of New York to the Acting Secretary of State, 15 March 1919, *Papers Relating to the Foreign Relations of the United States* [hereafter: *FRUS*], *1919, vol. II*, https://history.state.gov/historicaldocuments/frus1919v02/d198; Yuval Ben Bassat, 'Edoiut kartografiut mi-shelhai ha-tkofa ha-othmanit le-hipus neft ba-negev ve-bemidbar yehuda [Cartographic Evidence from the Late Ottoman Period of Oil Concession Permits in the Judean Desert and the Northern Negev]', *Ofakim Begeografia* [Horizons in Geography], 101–2 (2022), 128–9.
4. Novomeysky, *Given to Salt*, 15; Raja Shehadeh, *A Rift in Time: Travels with My Ottoman Uncle* (London: Profile Books, 2010), 52.
5. B.Z. Kedar, *Mabat ve'od mabat al eretz yisrael* [Looking Twice at the Land of Israel] (Jerusalem, 1991), 172; Album: 'World War I and the British Mandate in Palestine', Library of Congress, LOT 13827 (H), https://www.loc.gov/resource/ppmsca.13290/?sp=21&st=image.
6. Cited in: Michael Keren and Shlomit Keren, *We Are Coming, Unafraid: The Jewish Legions and the Promised Land in the First World War* (Lanham, MA: Rowman & Littlefield, 2010), 65.
7. Cited in: Eran Dolev, 'A Different Kind of Battle: Allenby's Anti-Malaria Campaign, Palestine, 1918', in E. Dolev, Y. Sheffy and H. Goren (eds), *Palestine and World War I* (London: I.B. Tauris, 2014), 133.
8. Matthew Hughes, 'Field Marshal Viscount Allenby: One of the Great Captains of History?', in Dolev, Sheffy and Goren (eds), *Palestine and World War I*, 113–15.
9. Major R.W. Brock, 'Potash from the Dead Sea' [no date – 1919], TNA, DSIR 26/58.
10. 'Report on the possible commercial utilisation of Dead Sea brine for the manufacture of potash and other salts', 1 November 1923, TNA, DSIR 26/58.
11. Yitzhak Gil-Har, 'Boundaries Delimitation: Palestine and Trans-Jordan', *Middle Eastern Studies*, 36:1 (2000), 72–3.

12. R.V. Vernon to Tulloch, 9 June 1922, TNA, CO 733/40/8.
13. Dov Gavish, *Melach Haaretz* [Salt of the Land] (Jerusalem, 1995), 100–1; Tal, *Mahapecha*, 19.
14. Novomeysky, *Given to Salt*, 255–62.
15. Maan Abu Nowar, *The Development of Trans-Jordan, 1929–1939* (Reading: Ithaca Press, 2006), 26; Norris, 'Toxic Waters', 36.
16. Alon, *The Making of Jordan*, 114; Yehuda Almog and Ben Zion Eshel, *Hevel Sdom* [The Region of Sodom] (Ein Harod, 1949), 93; Tal, *Mahapecha*, 142–3; Novomeysky, *Given to Salt*, 54.
17. Novomeysky, *Given to Salt*, 257; Tal, *Mahapecha*, 99, 170; J.M. Shapiro, Z. Saliternik and S. Belfekman, 'Malaria Survey of the Dead Sea Area during 1942, including the Description of a Mosquito Flight Test and its Results', *Transactions of the Royal Society of Tropical Medicine and Hygiene*, 38:2 (1944), 95–116; Alon Tal, *Pollution in a Promised Land* (University of California Press, 2002), 60; L. Naggan et al., 'Ecology and Attempted Control of Cutaneous Leishmaniasis around Jericho, in the Jordan Valley', *Journal of Infectious Diseases*, 121:4 (1970), 427.
18. Gavish, *Melach Haaretz*, 84–5; Tal, *Mahapecha*, 210–23, 262.
19. Sir E. John Russell, 'On Economizing Potash', *Agriculture*, 47:2 (1940), 109.
20. Tal, *Mahapecha*, 402; Norris, 'Toxic Waters', 35.
21. Gavish, *Melach Haaretz*, 100; David Koren, *Mol ha-yam ha-kaved: hevrat haashlag ve-ha-'Haganah'* [Facing the Heavy Sea: The Potash Company and the 'Haganah'] (Tel Aviv, 1991), 15.
22. Novomeysky, *Given to Salt*, 278–80.
23. Ben Gurion to Louis Brandeis, 'The Land Problem with Special Regard to Negev and Akaba', 4 June 1935, on display at the Ben Gurion Archives at Sde Boker.
24. Yair Giladi, *Haim hadashim le-yam ha-melach* [New Life for the Dead Sea] (Jerusalem, 1998), 102–5.
25. Koren, *Mol ha-yam ha-kaved*, 23–8, 34–5; 'Murder of Mr. G. Blake', *Palestine Post*, 5 July 1940, 2.
26. Petrucci to Ciano, 30 September 1940, *Documenti Diplomatici Italiani*, series 9, vol. V, 633–4; Yoav Gelber, *Toldot ha-hitnadvut* [The History of Volunteering], vol. I (Jerusalem: Yad Ben Zvi, 1979), 277–80.
27. 'Annex 4: Itinerary of the Special Committee in Palestine', https://www.un.org/unispal/document/auto-insert-186346/.
28. Gideon Biger, 'The Partition Plans for Palestine—1930–1947', *Israel Studies*, 26:3 (2021), 38–41.
29. Giladi, *Haim hadashim*, 121–6; Koren, *Mol ha-yam ha-kaved*, 61.
30. Koren, *Mol ha-yam ha-kaved*, 65–6.
31. Wells Stabler to the Acting Secretary of State, 4 December 1948, *FRUS, 1948, vol. V, part 2: The Near East, South Asia, and Africa*.
32. Justin McCarthy, *The Population of Palestine* (New York: Columbia University Press, 1990), 164, 167.
33. Atwa Jaber, 'No Bridge Will Take You Home: The Jordan Valley Exodus Remembered through the UNRWA Archives', *Jerusalem Quarterly*, 93 (2023), 10–32; Jalal Al Husseini, 'The Dilemmas of Local Development

and Palestine Refugee Integration in Jordan: UNRWA and the Arab Development Society in Jericho (1950–80)', *Jerusalem Quarterly*, 93 (2023), 61–79.
34. Confidential note by A.H. Wallace, December 1963, TNA, AVIA 120/5.
35. For reports in the Hebrew press on this incident, see: *Maariv*, 3 June 1954, 1; *Zmanim*, 4 June 1954, 1.
36. *Haboker*, 16 February 1955, 1.
37. Ronen Yitzhak, 'The Formation and Development of the Jordanian Air Force: 1948–1967', *Middle Eastern Studies*, 40:5 (2004), 165–6; Nir Hasson, 'Maagan avud, oniot atikot ve-matos yardeni [A lost anchorage, ancient boats and a Jordanian aircraft]', *Haaretz*, 26 November 2021, https://www.haaretz.co.il/science/archeology/2021-11-26/ty-article-magazine/.highlight/0000017f-db3c-df9c-a17f-ff3ceb690000.
38. Zafrir Rinat, 'ha-mishkaim be-araba ha-shanim ha-achronot gvohim me-hamemotza, ach yam ha-melach mamshich le-hitztamek [Precipitation in the last four years is higher than average, but the Dead Sea continues to shrink]', *Haaretz*, 25 December 2022, https://www.haaretz.co.il/nature/2022-12-25/ty-article/.premium/00000185-490e-df90-a797-cb0f71040001.
39. Sara Reguer, 'Rutenberg and the Jordan River: A Revolution in Hydro-Electricity', *Middle Eastern Studies*, 31:4 (1995), 691–729.
40. The conclusions of the water committee for the Jordan Valley kibbutzim, 14 February 1945, 'Haklaot [agriculture]', container 1-1-27/14, file 21178; Table summarising the amount of water used by the kibbutzim of the Jordan Valley, 18 December 1962, 'Haklaot [agriculture]', container 1-1-27/14, file 226, the Archive of Kibbutz Degania A; Courcier, Venot and Molle, 'Lower Jordan River Basin', 2, 11; Haddadin, 'A Jordanian Socio-Legal Perspective on Water Management', 42.
41. Tal, *Mahapecha*, 212–13; Shapiro, Saliternik and Belfekman, 'Malaria Survey', 95–116; Neev and Emery, 'The Dead Sea', 35–7.
42. Tal, *Pollution*, 208.
43. In 1944 the American engineer James B. Hayes, who had been involved in the Tennessee Valley Authority, became the chief engineer to the commission of experts working on developing Lowdermilk's plan. Hence, the subsequent work of the commission is often referred to as the Lowdermilk–Hayes project.
44. 'Plans for irrigation and hydro-electric development in Palestine', 3 August 1944, UN Archives, United Nations Relief and Rehabilitation Administration (UNRRA) (1943–1946).
45. Simcha Blass, *Mei meriva u-ma'as* [Water in Strife and Action] (Ramat Gan, 1973), 125–32; Donna Herzog, 'Contested Waterscapes: Constructing Israel's National Water Carrier' (PhD thesis, New York University, 2019), 26–7, 172–4, 180–1.
46. M. Klein and M. Halevi, 'Change in Water Discharge and Water Quality and its Effect on the Flora along the North Section of the Lower Jordan River', *Ofakim Begeografia* [Horizons in Geography], 22 (1987), 72; Neev and Emery, 'The Dead Sea', 35–7.
47. M.G. Ionides, 'The Possibilities of Irrigation Development in the Jordan Valley, Trans-Jordan', 7 July 1938, TNA, FO 816/97.

48. Office of the Historian, 'Historical Research Project no. 1403: The Jordan Valley Water Question 1919–1984', August 1984, iii–iv, 6; Frederic C. Hof, 'Dividing the Yarmouk's Waters: Jordan's Treaties with Syria and Israel', *Water Policy*, 1 (1998), 81–94.
49. Office of the Historian, 'Historical Research Project no. 1403', iv–v; Nathan J. Citino, 'The Ghosts of Development: The United States and Jordan's East Ghor Canal', *Cold War Studies*, 16:4 (2014), 159–88.
50. Abdul Mejid, 'The Jordan Valley', 26 August 1949, *The Times* (London), 5. See also: M.G. Ionides, 'The Disputed Waters of the Jordan', *Middle East Journal*, 17:2 (1953), 153–64.
51. Cited in: Citino, 'The Ghosts of Development', 173. For population figures see: P. Hindle, 'The Population of the Hashemite Kingdom of Jordan 1961', *The Geographical Journal*, 130:2 (1964), 261–4.
52. Citino, 'The Ghosts of Development', 184; Claud R. Sutcliffe, 'Palestinian Refugee Resettlement: Lessons from the East Ghor Canal Project', *Journal of Peace Research*, 11 (1974), 57–62.
53. Husseini, 'The Dilemmas of Local Development', 65.
54. Gabriel Stern, 'Nezirim etyopim be-kav ha-esh [Ethiopian monks in the line of fire]', *Al Ha-Mishmar*, 30 May 1969, 2.
55. *Davar*, 24 June 1969, 1; *Maariv*, 11 August 1969; *Maariv*, 11 September 1969, 1; Citino, 'The Ghosts of Development', 185–6; Jeffrey K. Sosland, *Cooperating Rivals: The Riparian Politics of the Jordan River Basin* (State University of New York Press, 2007), 98–101.
56. *Al-Ha-Mishmar*, 5 August 1977, 1; Office of the Historian, 'Historical Research Project no. 1403', 217–37.
57. Office of the Historian, 'Historical Research Project no. 1403', 237–8; Sosland, *Cooperating Rivals*, 108–14; Hof, 'Dividing the Yarmouk's Waters', 82; Courcier, Venot and Molle, 'Lower Jordan River Basin', 14.
58. Munther Haddadin, 'Water Issues in Hashemite Jordan', *Arab Studies Quarterly*, 22 (2000), 70; Ahmed A. al-Taani et al., 'Status of Water Quality in King Talal Reservoir Dam, Jordan', *Water Resources*, 45:4 (2018), 603–14.
59. Sosland, *Cooperating Rivals*, 182; Courcier, Venot and Molle, 'Lower Jordan River Basin', 18–21; Hana Namrouqa, 'Yarmouk water sharing violations require political solution', *The Jordan Times*, 28 April 2012.
60. Eliezer Schwarz et al., 'Yeridat miflas yam ha-melach: Teior, nitoach, hashlachot ve-pitronot [The Falling Level of the Dead Sea: Description, Analysis, Repercussions and Solutions]', Information and Research Center of the Knesset (Israeli parliament), 17 November 2008, 2.
61. Yuval Ne'eman and Israel Schul, 'Israel's Dead Sea Project', *Annual Review of Energy* (1983), 119.
62. Ibid., 128.
63. Ibid., 130.
64. Arab League Representatives to Kurt Waldheim, 20 October 1981, UN Archives, Secretary-General Kurt Waldheim collection (1972–1981).
65. Yehuda Zvi Blum to Kurt Waldheim, 2 October 1981, UN Archives, Secretary-General Kurt Waldheim collection (1972–1981).

NOTES TO PP. 197–203

66. UN General Assembly Resolution A/RES/37/122, 16 December 1982, https://research.un.org/en/docs/ga/quick/regular/37.
67. Raz, *Sefer Yam Ha-Melah*, 17.

CHAPTER 7 A SITE OF RUIN, BEAUTY, HEALTH AND HOPE

1. Abdelaziz Khlaifat et al., 'Mixing of Dead Sea and Red Sea Waters and Changes in their Physical Properties', *Heliyon*, 6:11 (2020); Ibrahim M. Oroud, 'The Future Fate of the Dead Sea: Total Disappearance or a Dwarfed Hypersaline Hot Lake?', *Journal of Hydrology*, 623 (2023). For historical and up-to-date Dead Sea level measurements, see the website of the Israeli Water Authority: https://www.gov.il/he/departments/publications/reports/dead_sea_report.
2. Nir Hasson, 'Kshe-koraim nevu'ot me-havar al atido shel yam ha-melach, kimat kashe lehaamin [When reading prophecies from the past on the future of the Dead Sea, it is almost hard to believe]', *Haaretz*, 29 December 2021, https://www.haaretz.co.il/magazine/2021-12-29/ty-article-magazine/.highlight/0000017f-e249-d804-ad7f-f3fb264e0000.
3. Orit Engelberg-Baram, 'Parashat Sdom: tmorut ba-yachas ha-zioni la-sviva derech mikreh mivchan meizamay pitoach shel yam-ha-melach be-sdom [The Sodom Affair: Changes in the Zionist Approach to the Environment through the Case Study of Development Initiatives of the Dead Sea at Sodom]', *Muza*, 5 (2022), 48.
4. 'Appraisal of the Dead Sea Works Ltd potash project', 7 June 1961; 'Report and recommendations of the President to the executive directors on a proposed loan to Mifalei Yam Hamelah B.M. (Dead Sea Works Limited)', 8 June 1961, World Bank Archives, https://projects.worldbank.org/en/projects-operations/document-detail/P037403.
5. Raz, *Sefer Yam Ha-Melah*, 65–6; Engelberg-Baram, 'Parashat Sdom', 42–69. For satellite imagery of the lake from c.1968, see CORONA 1104-2203Aft, https://corona.cast.uark.edu/.
6. Zafrir Rinat, 'Mifalei yam ha-melach choivu lishov mayim be-kafuf le-hayter, ach kamoyut ha-mayim ha-motarot rak gdelot [Dead Sea Works were obliged to pump water under authorization, but the quantity just keeps rising]', *Haaretz*, 27 May 2021; Zafrir Rinat, 'Matzavo shel yam ha-melach tzafuy le-hamir ekev harchavat pe'eilot ha-mifaalim she-betzidu ha-yardeni [The situation of the Dead Sea is set to worsen due to an expansion of activity in the plants on its Jordanian side]', *Haaretz*, 11 March 2024. For ICL's own estimate regarding their part in lowering the level of the lake, see: ICL, '2021 Corporate Responsibility (ESG) Report: Dead Sea Water Level', https://icl-group-sustainability.com/reports/dead-sea-water-level/.
7. Burton MacDonald, 'Aerial Photography and the Southern Ghors (al-Aghwār) and Northeast 'Arabah Archaeological Survey', *Studies in the History and Archaeology of Jordan*, 6 (1997), 95; David W. McCreery, 'Preliminary Report of the A.P.C. Township Archaeological Survey', *Annual of the Department of Antiquities of Jordan*, 22 (1977), 150–61; Omer Nawaf Maaitah and Salloom Al-Juboori, 'The Achievement of Safety Management

in Arab Potash Company', *Physical Sciences Research International*, 3:1 (2015), 1–7.
8. Rinat, 'Matzavo shel yam ha-melach'.
9. Neev and Emery, 'The Dead Sea', 60–61; Raz, *Sefer Yam Ha-Melah*, 83–8.
10. Raz, *Sefer Yam Ha-Melah*, 216.
11. Oren and Ventosa, 'Benjamin Elazari Volcani', 196–8.
12. 'Shitafon nitek kvish Sdom – Ein Gedi [Flood disconnects Sodom – Ein Gedi road]', *Ha-Boker*, 22 April 1964, 1.
13. J.L. Ashford and B.A. Boocock, 'Arab Potash Solar Evaporation System: Construction', *Proceedings of the Institution of Civil Engineers* (1984), 177–8.
14. Shai Gal's interview with Eli Raz in: 'Bol'anim be-yam ha-melach [Sinkholes in the Dead Sea]', *Channel 2 News*, 29 October 2012; Oriel Danielson's interview with Eli Raz in: 'Yam ha-melach kfi she-lo raitem oto me-olam [The Dead Sea like you've never seen before]', *Kan*, 1 November 2017; Engelberg-Baram, 'Parashat Sdom', 67.
15. 'Sinkholes hazard map along the Dead Sea coast', Israel Geological Survey, 9 February 2020, https://www.gov.il/en/departments/general/feasibility-to-create-sinkholes.
16. Moshe Gilad, 'Mad ha-chom eeyem le-hitpotzetz [The thermometer was ready to explode]', *Haaretz*, 27 August 2019, https://www.haaretz.co.il/gallery/trip/2019-08-27/ty-article-magazine/.premium/0000017f-dbb4-db22-a17f-ffb568df0000.
17. Salah Taqieddin, Nabil S. Abderahman and Mohammad Atallah, 'Sinkhole Hazards along the Eastern Dead Sea Shoreline Area, Jordan: A Geological and Geotechnical Consideration', *Environmental Earth Sciences*, 39:11 (2000), 1237–53; Damien Closson et al., 'Subsidence and Sinkhole Hazards in the Dead Sea Area, Jordan', *Pure and Applied Geophysics*, 162 (2005), 223.
18. Synne Furnes Bjerkestrand, 'Earth Day: Jordan farmers frustrated over shrinking Dead Sea', *Al-Jazeera*, 21 April 2022, https://www.aljazeera.com/news/2022/4/21/earth-day-jordan-farmers-frustrated-over-shrinking-of-dead-sea; Saeb Rawashdeh, 'Dead Sea rebounds from pandemic slump as tourist magnet for Jordan', *The Circuit*, 18 December 2022, https://circuit.news/2022/12/18/dead-sea-rebounds-from-pandemic-slump-as-tourist-magnet-for-jordan/.
19. Moshe Gilad, 'U-bein ha-time be-yam ha-melach: ha-karka nishmetet tachat ha-raglayim [Meanwhile in the Dead Sea: The ground drops from under the feet]', *Haaretz*, 18 December 2023; Eyal Shalev (ed.), 'Tgovat ha-tashtit be-agan yam ha-melach le-shinoyim ha-tive'yim ve-maase yedei adam [The response of infrastructure in the Dead Sea basin to natural and man-made changes]', Geological Survey of Israel, January 2022; the author's interview with Moshe Gilad, 20 December 2023.
20. 'Ha-nahar ha-sodi shel Israel [Israel's secret river]', *Kan*, 20 January 2020, https://www.facebook.com/watch/?v=621188132016733.
21. 'Siyum hakirat ason Ein Gedi [End of enquiry of the Ein Gedi tragedy]', *Haaretz*, 17 May 1942, 4.
22. Ilan Troen and Shay Rabineau, 'Competing Concepts of Land in Eretz Israel', *Israel Studies*, 19:2 (2014), 178; Moshe Gilad, 'Hazarti la-shvil

ha-mesoman ha-rishon baaretz [I came back to the first marked trail in the country]', *Haaretz*, 5 June 2017; Shay Rabineau, *Walking the Land: A History of Israeli Hiking Trails* (Bloomington: Indiana University Press, 2023), 90–101.
23. Azaria Alon, 'Ein Gedi ke-shmorat teva [Ein Gedi as a nature reserve]', *La-Merhav*, 6 March 1964, 4.
24. *Ha-Tzofeh*, 22 May 1968, 4; *Al Ha-Mishmar*, 20 October 1969, 2.
25. Naif A. Haddad, Sharaf A. al-Kheder and Leen A. Fakhoury, 'Al Mujib Natural Reserve in Jordan: Towards an Assessment for Sustainable Ecotourism Management Plan Utilizing Spatial Documentation', *Natural Resources and Conservation*, 1:3 (2013), 65–76.
26. 'King Abdullah's grief turns to anger over Jordanian flood deaths', *The National*, 26 October 2018.
27. Rafael Schäffer and Ingo Sass, 'The Thermal Springs of Jordan', *Environmental Earth Sciences*, 72 (2014), 171–87.
28. Ada Schein and Dov Gavish, 'Briut ve-yofi ke-manof le-pitoah hevel yam ha-melah [Health and Beauty as an Impetus to the Development of the Dead Sea Region]', *Karka*, 66 (2009), 86–128.
29. 'Tvilat botz tiv'it be-Ein Bokek [A natural mud bath in Ein Bokek]', *Haaretz*, 13 November 1961, 5; Dov Gavish and Ada Schein, *Sipuro shel mifal ha-tamrukim le-hof yam ha-melach* [The Story of the Cosmetics Factory on the Shore of the Dead Sea] (Jerusalem: Ariel, 2007), 35–6.
30. Tal, *Mahapecha*, 262.
31. B. Adler, 'Aizor yam ha-melach yahafuch le-merkaz bein-leumi le-tayarut marpe [The Dead Sea region will become an international centre for health tourism]', *Ha-Tzofe*, 6 October 1968, 7; David Moshiov, 'Shiluv shel tayarut nofesh ve-marpe [A combination of leisure and health tourism]', *Davar*, 17 July 1981, 57.
32. Shimon Moses, 'Yam ha-melach: atar tivai le-marpe u-lebriut [The Dead Sea: A Site of Convalescence and Health]', in Oded Navon (ed.), *Melach Haaretz* [Salt of the Land], vol. 2 (Jerusalem, 2006), 98–107.
33. Schein and Gavish, 'Briut ve-yofi', 86–128.
34. 'Boycott of Ahava Dead Sea products makes an impact', BDS movement website, 3 December 2009, https://bdsmovement.net/news/boycott-ahava-dead-sea-products-makes-impact.
35. Nancy Oster, 'Ahava London flagship to close over demonstrations', *JTA*, 21 September 2011.
36. 'Cosmetics giant Ahava opens plant on Israeli side of green line', *Times of Israel*, 11 March 2016.
37. Eli Alon, 'Matzevet zikaron la-tayas ha-yardeni be-Ein Gedi [Memorial for the Jordanian pilot in Ein Gedi]', *News1*, 17 November 2021, https://www.news1.co.il/Archive/0026-D-151362-00.html.
38. Khairieh 'Amr et al., 'Archaeological Survey of the East Coast of the Dead Sea Phase I: Suwayma, Az-Zara and Umm Sidra', *Annual of the Department of Antiquities of Jordan*, 40 (1996), 429–49.
39. Patricia M. Bikai and Virginia Egan, 'Archaeology in Jordan', *American Journal of Archaeology*, 100:3 (1996), 507–35; Mohammad Waheeb, 'The

Discovery of Bethany Beyond the Jordan River (Wadi Al-Kharrar)', *Dirasat, Human and Social Sciences*, 35:1 (2008), 115–26; Naif Haddad et al., 'The Baptism Archaeological Site of Bethany Beyond Jordan', *Tourism and Hospitality Planning & Development*, 6:3 (2009), 173–90.
40. 'Jordan plans to attract 1 million visitors to revamped Jesus baptism site in 2030', *The New Arab*, 16 January 2023.
41. 'Sant Francisco al ha-mayim [St Francis on the water]', *Kol Ha-Ir*, 30 October 1987, 19.
42. Oroud, 'The Future Fate'. For other estimates, see: Raz, *Sefer Yam Ha-Melah*, 73.
43. '"It feels really natural": hundreds pose nude for Spencer Tunick shoot near Dead Sea', *Guardian*, 18 October 2021.
44. The author's interview with Oded Rahav, May 2024.
45. John Anthony Allan, Abdallah I. Husein Malkawi and Yacov Tsur, 'Red Sea–Dead Sea Water Conveyance Study Program: Final Report', March 2014.
46. Royal HaskoningDHV, 'Red Sea Dead Sea Water Conveyance Study', submitted to the European Investment Bank, 6 December 2017.
47. The author's interview with Leehee Goldenberg, 7 September 2022; Michael Beyth, 'The Red Sea and the Mediterranean–Dead Sea Canal Project', *Desalination*, 214 (2007), 364–70.
48. 'World Bank releases statement on Jordan's "Red Sea–Dead Sea" project', *Roya News*, 29 May 2021.
49. EcoPeace, 'EcoPeace's Vision for Stabilizing the Water Level of the Dead Sea', 2018.
50. Israel's Ministry of Environmental Protection, 'Joining forces to address climate change impact: Israel and Jordan to cooperate on Jordan River restoration', 17 November 2002, https://www.gov.il/en/departments/news/israel_jordan_environmental_cooperation.
51. Eli Raz, 'The Future of the Dead Sea: Is the Red Sea–Dead Sea Conduit the Right Solution?', in Lipchin, Sandler and Cushman (eds), *The Jordan River and Dead Sea Basin*, 191.
52. Nir Hasson, 'Ze proyekt anak ve-shanuyi be-machluket be-drom ha-aaretz. Ba-misrad le-hagant ha-sviva mityazvim meachorav [It's a huge and controversial project in the south of the country. At the Environment Ministry they stand behind it]', *Haaretz*, 20 June 2022.

EPILOGUE: A HITCHHIKER'S GUIDE TO THE BELEAGUERED DEAD SEA

1. Amanda Borschel-Dan, 'What matters now to Dr. Joe Uziel: ID'ing Oct. 7 victims via forensic archaeology', *Times of Israel*, 17 November 2023, https://www.timesofisrael.com/what-matters-now-to-dr-joe-uziel-iding-oct-7-victims-via-forensic-archaeology/.
2. 'IDF shows remains of Iranian ballistic missile that fell in Dead Sea', *Times of Israel*, 17 April 2024.

3. See, for instance: 'UNESCO votes to list ruins near ancient Jericho as a World Heritage Site in Palestine', *Times of Israel*, 17 September 2023, https://www.timesofisrael.com/un-committee-votes-to-list-ancient-jericho-as-a-world-heritage-site-in-palestine/.
4. Carl Alpert, 'Peace on Israel', *Canadian Jewish Chronicle Review*, 8 December 1967, 10.
5. 'Israel to authorize new Jordan Valley settlement after annexation pledge', *Jerusalem Post*, 12 September 2019.
6. Nir Hasson, 'Sreifat mata ha-tmarim be-Almog kilta alfei etzim ve-libta gam metachim pnimyim [The burning of the date palm grove in Almog consumed thousands of trees and also stoked internal tensions]', *Haaretz*, 2 June 2023.
7. 'Amr et al., 'Archaeological Survey', 448; Bikai and Egan, 'Archaeology in Jordan', 575–6.
8. The author's interview with Moshe Gilad, 20 December 2023; Allyssia Alleyne, 'Sigalit Landau's gown of salt crystals rises from Dead Sea', *CNN*, 15 November 2016.
9. Sue Surkes, 'Authorities urged to act immediately to stabilize Dead Sea levels', *Times of Israel*, 20 June 2022.
10. Zafrir Rinat, 'Hitkavzut yam ha-melach lo ne'ezeret ve-ha-hitmodedut im ha-hashlachot holechet ve-mistabechet [The shrinking of the Dead Sea is not stopping and the responses to its consequences are becoming more complex]', *Haaretz*, 17 April 2024.
11. Tristram, *Land of Israel*, 336.
12. Adi Wolfson, 'David mul Goliath, girsat yam ha-melach ve-hasviva [David vs Goliath, a version [featuring] the Dead Sea and the environment]', *Ynet*, 17 December 2022.

SELECT BIBLIOGRAPHY

THE GEOLOGY OF THE JORDAN VALLEY AND THE DEAD SEA

Abu Ghazleh, Shahrazad. 'Lake Lisan and the Dead Sea: Their Level Changes and the Geomorphology of their Terraces' (PhD thesis, Technische Universität Darmstadt, 2011)

Agnon, Amotz. 'Pre-Instrumental Earthquakes along the Dead Sea Rift', in Z. Garfunkel et al. (eds), *Dead Sea Transform Fault System: Reviews* (Dordrecht: Springer, 2014)

Bandel, Klaus, Ikhlas Alhejoj and Elias Salameh. 'Geologic Evolution of the Tertiary-Quaternary Jordan Valley with Introduction of the Bakura Formation', *Freiberger Forschungshefte*, 23 (2016), 103–35

Bartov, Yuval et al. 'Lake Levels and Sequence Stratigraphy of Lake Lisan, the Late Pleistocene Precursor of the Dead Sea', *Quaternary Research*, 57:1 (2002)

Freund, Raphael. 'A Model of the Structural Development of Israel and Adjacent Areas since Upper Cretaceous Times', *Geological Magazine*, 102:3 (1965)

Garfunkel, Zvi. 'Ha-tektonika shel transform yam ha-melach [The Tectonics of the Dead Sea Transform]', *Melach Haaretz*, 2 (2006)

Horowitz, Aharon et al. *The Jordan Rift Valley* (Lisse: Balkema, 2001)

Neev, David and Kenneth Orris Emery. 'The Dead Sea: Depositional Process and Environments of Evaporates', *Geological Survey of Israel*, bulletin no. 41 (1967)

Quennell, Albert Mathieson. 'The Structural and Geomorphic Evolution of the Dead Sea Rift', *Quarterly Journal of the Geological Society of London*, 64 (1958)

Raz, Eli. *Sefer Yam Ha-Melah* [Book of the Dead Sea] (Jerusalem: Nature and Parks Authority Press, 1993)

Waldmann, Nicolas et al. 'Stratigraphy, Depositional Environments and Level Reconstruction of the Last Interglacial Lake Samra in the Dead Sea Basin', *Quaternary Research*, 72 (2009)

ANCIENT CLIMATE AND CHANGES IN THE LEVEL OF THE DEAD SEA

Bookman, R. et al. 'Late Holocene Lake Levels of the Dead Sea', *Geological Society of America Bulletin*, 116 (2004)

SELECT BIBLIOGRAPHY

Kagan, Elisa Joy et al. 'Dead Sea Levels during the Bronze and Iron Ages', *Radiocarbon*, 57:2 (2015)

Levi, Yehuda et al. 'Harnessing Paleohydrologic Modeling to Solve a Prehistoric Mystery', *Scientific Reports*, 9 (2019)

Litt, Thomas et al. 'Holocene Climate Variability in the Levant from the Dead Sea Pollen Record', *Quaternary Science Reviews*, 49 (2012)

Migowski, Claudia et al. 'Holocene Climate Variability and Cultural Evolution in the Near East from the Dead Sea Sedimentary Record', *Quaternary Research*, 66 (2006)

Neumann, Frank H. et al. 'Vegetation History and Climate Fluctuations on a Transect along the Dead Sea West Shore and their Impact on Past Societies over the Last 3500 Years', *Journal of Arid Environments*, 74:7 (2010)

Stein, Mordechai et al. 'Abrupt Aridities and Salt Deposition in the Post-Glacial Dead Sea and their North Atlantic Connection', *Quaternary Science Reviews*, 29 (2010)

Weber, Nurit et al. 'Gypsum Deltas at the Holocene Dead Sea Linked to Grand Solar Minima', *Geophysical Research Letters* (2021)

THE ARCHAEOLOGY OF THE REGION FROM THE NATUFIANS TO THE IRON AGE

'Amr, Khairieh et al. 'Archaeological Survey of the East Coast of the Dead Sea Phase I: Suwayma, Az-Zara and Umm Sidra', *Annual of the Department of Antiquities of Jordan*, 40 (1996)

Bar-Yosef, Ofer. 'The Natufian Culture in the Levant, Threshold to the Origins of Agriculture', *Evolutionary Anthropology*, 6:5 (1998)

Bar-Yosef, Ofer. 'The Walls of Jericho: An Alternative Interpretation', *Current Anthropology*, 27:2 (1986)

Barkai, Ran and Roy Liran. 'Midsummer Sunset at Neolithic Jericho', *Time and Mind*, 1:3 (2008)

Belfer-Cohen, Anna. 'The Natufian in the Levant', *Annual Review of Anthropology*, 20 (1991)

Bunch, Ted et al. 'A Tunguska Sized Airburst Destroyed Tall el-Hammam, a Middle Bronze Age City in the Jordan Valley near the Dead Sea', *Scientific Reports*, 11:1 (2021)

Drabsch, Bernadette. 'The Wall Art of Teleilat Ghassul: When, Where, Why, To Whom and by Whom?', *Expression*, 8 (2015)

Edwards, Phillip C. 'Natufian Interactions along the Jordan Valley', *Palestine Exploration Quarterly*, 147:4 (2015)

Finkelstein, Israel. 'Jeroboam II in Transjordan', *Scandinavian Journal of the Old Testament*, 34:1 (2020)

Gallo, Elisabetta. 'Tell es-Sultan/Ancient Jericho in the Early Bronze Age II-III: A Population Estimate', *ROSAPAT*, 13 (2019)

Hamer, Hagay et al. 'The Neolithic Occupation of the Judean Desert Caves in Light of Recent Surveys and Excavations on the Eastern Cliffs', in *New Studies in the Archaeology of the Judean Desert* (Jerusalem: Israel Antiquities Authority, 2023)

Ji, Chang-Ho. 'A Moabite Sanctuary at Khirbat Ataruz, Jordan: Stratigraphy, Findings, and Archaeological Implications', *Levant*, 50:2 (2018)

Kaptijn, Eva. *Life on the Watershed* (Leiden: Sidestone Press, 2009)

Kenyon, Kathleen M. 'Excavations at Jericho', *The Journal of the Royal Anthropological Institute of Great Britain and Ireland*, 84:1/2 (1954)

Kuijt, Ian et al. 'Pottery Neolithic Landscape Modification at Dhra'', *Antiquity*, 81 (2007)

SELECT BIBLIOGRAPHY

Langgut, Dafna and Yosef Garfinkel. '7000-Year-Old Evidence of Fruit Tree Cultivation in the Jordan Valley, Israel', *Scientific Reports*, 12:7463 (2022)

Mithen, Steven. *After the Ice: A Global History 20,000–5,000 BC* (London: Weidenfeld & Nicolson, 2003)

Nigro, Lorenzo. 'Results of the Italian-Palestinian Expedition to Tell es-Sultan: At the Dawn of Urbanization in Palestine', *ROSAPAT*, 2 (2006)

Nigro, Lorenzo. 'The Archaeology of Collapse and Resilience: Tell es-Sultan/Ancient Jericho as a Case Study', *ROSAPAT*, 11 (2014)

Nigro, Lorenzo. 'The Italian-Palestinian Expedition to Tell es-Sultan, Ancient Jericho (1997–2015): Archaeology and Valorisation of Material and Immaterial Heritage', in Rachael Thyrza Sparks et al. (eds), *Digging Up Jericho: Past, Present and Future* (Oxford: Archaeopress, 2020)

Peilstöcker, Martin and Sabine Wolfram (eds). *Life at the Dead Sea: Proceedings of the International Conference held at the State Museum of Archaeology Chemnitz (smac), February 21–24, 2018, Chemnitz* (Münster: Zaphon, 2019)

Politis, Konstantinos D. (ed.). *Ancient Landscapes of Zoara I* (London: Routledge, 2020)

Rainey, Anson F. 'Shasu or Habiru: Who Were the Early Israelites?', *Biblical Archaeology Review*, 34:6 (2008)

Rast, Walter E. and Thomas Schaub. 'Survey of the Southeast Plain of the Dead Sea', *Annual of the Department of Antiquities of Jordan*, 19 (1973)

Ripepi, Gaia. 'Mudbricks and Modular Architecture at Tell es-Sultan from the Neolithic to the Bronze Age', in Rachael Thyrza Sparks et al. (eds), *Digging Up Jericho: Past, Present and Future* (Oxford: Archaeopress, 2020)

Schick, Tamar. *The Cave of the Warrior: A Fourth Millennium Burial in the Judean Desert* (Jerusalem: Israel Antiquities Authority, 1998)

Shqiarat, Mansour. 'History and Archaeology of Water Management in Jordan', *Scientific Culture*, 5:3 (2019)

Ussishkin, David. 'The Chalcolithic Temple in Ein Gedi: Fifty Years after its Discovery', *Near Eastern Archaeology*, 77:1 (2014)

THE DEAD SEA IN THE HELLENISTIC, HASMONEAN, ROMAN AND BYZANTINE PERIODS

Bianchi, Davide. 'New Archaeological Discoveries in the Basilica of the Memorial of Moses, Mount Nebo', *Studies in the History and Archaeology of Jordan*, 13 (2019)

Binder, Stéphanie E., Michael Lazar and Emmanuel Nantet. 'Measurements and Shape of the Dead Sea in the Hellenistic and Roman Periods', *Israel Exploration Journal*, 69:2 (2019)

Cotton, Hannah. 'Ein Gedi Between the Two Revolts', *Scripta Classica Israelica*, 20 (2001)

Cotton, Hannah. 'The Date of the Fall of Masada: The Evidence of the Masada Papyri', *Zeitschrift für Papyrologie und Epigraphik*, 78 (1989)

Friedman, Zaraza. 'Sailing in the Dead Sea: Madaba Map Mosaic', in L. Daniel Chrupcala (ed.), *Christ Is Here! Studies in Biblical and Christian Archaeology in Memory of Michele Piccirillo* (Milano: Edizioni Terra Santa, 2012)

Garbrecht, Günter and Yehuda Peleg. 'The Water Supply of the Desert Fortresses in the Jordan Valley', *The Biblical Archaeologist*, 57:3 (1994)

Hadas, Gideon. 'The Balsam "Afarsemon" and Ein Gedi during the Roman-Byzantine Period', *Revue Biblique* (2007)

Hadas, Gideon, Nili Liphschitz and Georges Bonani. 'Two Ancient Wooden Anchors from Ein Gedi, on the Dead Sea, Israel', *International Journal of Nautical Archaeology*, 34:2 (2005)

SELECT BIBLIOGRAPHY

Haddad, Naif et al. 'The Baptism Archaeological Site of Bethany Beyond Jordan', *Tourism and Hospitality Planning & Development*, 6:3 (2009)

Hepper, Nigel F. and Joan E. Taylor. 'Date Palms and Opobalsam in the Madaba Mosaic Map', *Palestine Exploration Quarterly*, 136:1 (2004)

Hirschfeld, Yizhar. 'A Climatic Change in the Early Byzantine Period? Some Archaeological Evidence', *Palestine Exploration Quarterly*, 136:2 (2004)

Hirschfeld, Yizhar and Donald T. Ariel. 'A Coin Assemblage from the Reign of Alexander Jannaeus Found on the Shore of the Dead Sea', *Israel Exploration Journal*, 55:1 (2005)

Izdebski, Adam. 'Why Did Agriculture Flourish in the Late Antique East?', *Millennium*, 8:1 (2011)

Klein, Eitan et al. 'A Hoard of Ptolemaic Coins found in Murabba'at Cave II in the Judean Desert', *Atiqot*, 112 (2023)

Leal, Beatrice. 'A Reconsideration of the Madaba Map', *Gesta*, 57:2 (2018)

Magness, Jodi. *Masada: From Jewish Revolt to Modern Myth* (Princeton University Press, 2019)

Magness, Jodi. *The Archaeology of the Holy Land* (Cambridge University Press, 2012)

Manolaraki, Eleni. 'Hebraei Liquores: The Balsam of Judaea in Pliny's Natural History', *The American Journal of Philology*, 136:4 (2015)

Mayerson, Philip. 'The Saracens and the *Limes*', *Bulletin of the American Schools of Oriental Research*, 262 (1986)

Meshel, Ze'ev. 'The Nabataean "Rock" and the Judaean Desert Fortresses', *Israel Exploration Journal*, 50:1/2 (2000)

Netzer, Ehud. *The Architecture of Herod, the Great Builder* (Tübingen: Mohr Siebeck, 2020)

Nissenbaum, A. 'Dead Sea Asphalt in Ancient Egyptian Mummies - Why?', *Archaeometry*, 55:3 (2013)

Oron, Asaf et al. 'Two Artificial Anchorages off the Northern Shore of the Dead Sea', *International Journal of Nautical Archaeology*, 44:1 (2015)

Rullkötter, J. and A. Nissenbaum. 'Dead Sea Asphalt in Egyptian Mummies: Molecular Evidence', *Naturwissenschaften*, 75 (1988)

Sallon, Sarah et al. 'Origins and Insights into the Historic Judean Date Palm based on Genetic Analysis of Germinated Ancient Seeds and Morphometric Studies', *Science Advances*, 6 (2020)

Sivan, Hagith. *Palestine in Late Antiquity* (Oxford University Press, 2008)

Taylor, Jane. *Petra and the Lost Kingdom of the Nabataeans* (London: I.B. Tauris, 2001)

Vörös, Győző. 'Machaerus: The Golgotha of John the Baptist', *Revue Biblique*, 119:2 (2012)

Waheeb, Mohammad. 'The Discovery of Bethany Beyond the Jordan River (Wadi Al-Kharrar)', *Dirasat, Human and Social Sciences*, 35:1 (2008)

THE DEAD SEA SCROLLS AND THE SEARCH FOR ANCIENT TEXTS IN THE JUDEAN DESERT

Anava, Sarit et al. 'Illuminating Genetic Mysteries of the Dead Sea Scrolls', *Cell*, 181 (2020)

Dershowitz, Idan. 'The Valediction of Moses: New Evidence on the Shapira Deuteronomy Fragments', *Zeitschrift für die alttestamentliche Wissenschaft*, 133:1 (2021)

Dušek, Jan. 'The Importance of the Wadi Daliyeh Manuscripts for the History of Samaria and the Samaritans', *Religions*, 11:63 (2020)

SELECT BIBLIOGRAPHY

Eshel, Hanan and Boaz Zissu. *The Bar Kokhba Revolt: The Archaeological Evidence* (Jerusalem: Yad Ben Zvi, 2019)
Fields, Weston. *The Dead Sea Scrolls: A Full History, 1947–1960* (Leiden: Brill, 2009)
Golb, Norman. 'Khirbet Qumran and the Manuscripts of the Judaean Wilderness: Observations on the Logic of their Investigation', *Journal of Near Eastern Studies*, 49:2 (1990)
Graystone, G. 'The Dead Sea Scrolls - II. Wadi Murabba'at', *Scripture*, 7:3 (1955)
Guil, Shlomo. 'The Shapira Scroll was an Authentic Dead Sea Scroll', *Palestine Exploration Quarterly*, 149:1 (2017)
Lim, Timothy H. *The Dead Sea Scrolls: A Very Short Introduction* (Oxford University Press, 2017)
Talmon, Shemaryahu. 'Hebrew Written Fragments from Masada', *Dead Sea Discoveries*, 3:2 (1996)
Taylor, Joan. *The Essenes, the Scrolls, and the Dead Sea* (Oxford University Press, 2012)
Vainstub, Daniel. 'The Covenant Renewal Ceremony as the Main Function of Qumran', *Religion*, 12:578 (2021)
de Vaux, Roland. *The Dead Sea Scrolls, A Full History* (Leiden: Brill, 2009)
Vermès, Géza. *Scripture and Tradition in Judaism: Haggadic Studies* (Leiden: Brill, 1983)
Yadin, Yigael. *Bar Kokhba: The Rediscovery of the Legendary Hero of the Last Jewish Revolt Against Imperial Rome* (London: Weidenfeld and Nicolson, 1971)

THE REGION DURING THE MEDIEVAL AND EARLY MODERN PERIODS

Amar, Zohar. 'The Production of Salt and Sulphur from the Dead Sea Region in the Tenth Century according to at-Tamimi', *Palestine Exploration Quarterly*, 130 (1998)
Arad, Pnina. 'Landscape and Iconicity: Proskynetaria of the Holy Land from the Ottoman Period', *The Art Bulletin*, 100:4 (2018)
Ashtor, Eliyahu. *Technology, Industry and Trade: The Levant versus Europe, 1250–1500* (Ashgate: Routledge, 1992)
Avni, Gideon. *The Byzantine–Islamic Transition in Palestine* (Oxford University Press, 2014)
Bailey, Clinton. 'Dating the Arrival of Bedouin Tribes in Sinai and the Negev', *Journal of the Economic and Social History of the Orient*, 28:1 (1985)
Bronstein, Judith, Edna J. Stern and Elisabeth Yehuda. 'Franks, Locals and Sugar Cane: A Case Study of Cultural Interaction in the Latin Kingdom of Jerusalem', *Journal of Medieval History*, 45:3 (2019)
Ellenblum, Ronnie. *Frankish Rural Settlements in the Latin Kingdom of Jerusalem* (Cambridge University Press, 1998)
Fulcher of Chartres. *A History of the Expedition to Jerusalem 1095–1127* (University of Tennessee Press, 1969)
Fuller, Thomas. *A Pisgah Sight of Palestine and the Confines Thereof* (London: John Williams, 1650)
Hasselquist, Fredrik. *Voyages and Travels in the Levant in the Years 1749, 50, 51, 52* (London, 1766)
Hütteroth, Wolf-Dieter and Kamal Abdulfattah. *Historical Geography of Palestine, Transjordan and Southern Syria in the Late 16th Century* (Erlangen, 1977)
Istakhri, I. *al-Masalik wa al-Mamalik* [Roads and Kingdoms] (Beirut, 2004)
Le Strange, Guy. *Palestine under the Moslems* (London: Alexander P. Watt, 1890)
Morton, Nicholas. *The Mongol Storm* (London: Basic Books, 2023)
Nissenbaum, Arie. 'The Dead Sea Monster', *International Journal of Salt Lake Research*, 1:2 (1992)

SELECT BIBLIOGRAPHY

Pococke, Richard. *A Description of the East and Some Other Countries* (London, 1745)
Politis, Konstantinos D. 'The Sugar Industry in the Ghawr as-Safi, Jordan', *Studies in the History and Archaeology of Jordan*, 11 (2013)
Pringle, Denys. *Pilgrimage to Jerusalem and the Holy Land, 1187–1291* (London: Routledge, 2012)
Pringle, Denys. 'Templar Castles on the Road to the Jordan', in Malcolm Barber (ed.), *The Military Orders: Fighting for the Faith and Caring for the Sick* (Ashgate: Routledge, 1994), 148–66
Proskynētaria tōn Hagiōn Topōn: deka hellēnika cheirographa 16ou–18ou ai [Pilgrimages of the Holy Places: Ten Greek manuscripts 16th–18th centuries] (Thessaloniki, 1986)
Rubin, Rehav. 'Greek-Orthodox Maps of Jerusalem from the Seventeenth and Eighteenth Centuries', *e-Perimetron*, 8:3 (2013)
Schenk, Jochen G. 'Nomadic Violence in the First Latin Kingdom of Jerusalem and the Military Orders', *Reading Medieval Studies*, 36 (2010)
Taha, Hamdan. 'The Sugarcane Industry in Jericho, Jordan Valley', in Konstantinos D. Politis (ed.), *The Origins of the Sugar Industry and the Transmission of Ancient Greek and Medieval Arab Science and Technology from the Near East to Europe* (Athens: National and Kapodistriako University of Athens, 2015), 51–77
Taylor, Joan. 'The Dead Sea in Western Travellers' Accounts from the Byzantine to the Modern Period', *Strata*, 27 (2009), 9–30
de Vitry, Jacques. *History of Jerusalem*, translated by Aubrey Stewart (London: Palestine Pilgrims' Text Society, 1896)
Walker, Bethany J. 'Mamluk Investment in Southern Bilād Al Shām in the Eighth/ Fourteenth Century: The Case of Ḥisbān', *Journal of Near Eastern Studies*, 62:4 (2003)
John of Würzburg. *Description of the Holy Land* (London, 1890)

NINETEENTH- AND EARLY-TWENTIETH-CENTURY EXPLORATION

Allen, William. *The Dead Sea: A New Route to India*, vol. I (London: Longman, 1855)
de Bertou, Jules. 'Notes on a Journey from Jerusalem by Hebron, the Dead Sea, El Ghór, and Wádí Arabah to Aḳabah, and Back by Petra; in April, 1838', *The Journal of the Royal Geographical Society of London*, 9 (1839)
Conder, C.R. and H.H. Kitchener. *The Survey of Western Palestine: Memoirs of the Topography, Orography, Hydrography and Archaeology* (London, 1883)
Forder, Archibald. *With the Arabs in Tent and Town* (London: Marshal Brothers, 1902)
Goren, Haim. *Dead Sea Level: Science, Exploration and Imperial Interests in the Middle East* (London: I.B. Tauris, 2011)
Herzl, Theodor. *Old New Land*, translated by Lotta Levensohn (Princeton: Markus Wiener, 2007)
Hill, John Gray. 'The Dead Sea', *Palestine Exploration Fund Quarterly Statement* (1900)
Hudleston, Wilfrid H. *The Geology of Palestine* (London, 1885)
Hull, Edward. *The Survey of Western Palestine: Memoir on the Physical Geology and Geography of Arabia Petraea, Palestine, and Adjoining Districts* (London, 1886)
Huntington, Ellsworth. *Palestine and its Transformation* (Boston and New York: Houghton Mifflin, 1911)
International River Jordan Water Company, *Souvenir Album of the Holy Land with Exceptionally Rare Views of the Sacred River Jordan* (New York, 1907)
Kreiger, Barbara. *Living Waters: Myth, History and Politics of the Dead Sea* (New York: Continuum, 1988)
Lartet, Louis. *Exploration géologique de la mer Morte de la Palestine et de l'Idumée* (Paris: Bertrand, 1877)

SELECT BIBLIOGRAPHY

Lortet, Louis. *La Syrie d'aujourd'hui: voyages dans la Phénicie, le Liban, et la Judée, 1875–1880* (Paris, 1884)

Lortet, Louis. 'Researches on the Pathogenic Microbes of the Mud of the Dead Sea', *Palestine Exploration Quarterly*, 24:1 (1892)

Lynch, William Francis. *Narrative of the United States' Expedition to the River Jordan and the Dead Sea* (Philadelphia, 1850)

Masterman, E.W.G. 'Summary of the Observations on the Rise and Fall of the Level of the Dead Sea, 1900–1913', *Palestine Exploration Quarterly*, 45:4 (1913)

Pappalardo, Carmelo. 'Una lettera di Paolo Bajnotti a Cristoforo Negri nel contesto dell'esplorazione del Mar Morto nel XIX secolo', *Bollettino della Società Geografica Italiana*, 14:4 (2021)

Press, Michael. 'The French Connection: Charles Clermont Ganneau and the Antiquities Trade Network in the Late Ottoman Levant', *Journal of Ancient Judaism*, 14 (2023)

Robinson, Edward. *Biblical Researches in Palestine, Mount Sinai and Arabia Petraea*, vol. II (Boston: Crocker & Brewster, 1841)

Seetzen, Ulrich Jasper. *Reisen durch Syrien, Palästina, Phönicien, die Transjordan-länder, Arabia Petraea und Unter-Aegypten* (Berlin, 1859)

Tristram, Henry Baker. *Bible Places; or, The Topography of the Holy Land* (London: Society for Promoting Christian Knowledge, 1884)

Tristram, Henry Baker. *The Land of Israel: A Journal of Travels in Palestine* (London, 1865)

Twain, Mark. *The Innocents Abroad* (Hartford: American Publishing Company, 1869)

THE DEAD SEA AND JORDAN VALLEY SINCE THE FIRST WORLD WAR

Abu Nowar, Maan. *The Development of Trans-Jordan, 1929–1939* (Reading: Ithaca Press, 2006)

Biger, Gideon. 'The Partition Plans for Palestine—1930–1947', *Israel Studies*, 26:3 (2021)

Blass, Simcha. *Mei meriva u-ma'as* [Water in Strife and Action] (Ramat Gan, 1973)

Citino, Nathan J. 'The Ghosts of Development: The United States and Jordan's East Ghor Canal', *Cold War Studies*, 16:4 (2014)

Cobbing, Felicity. 'John Garstang's Excavations at Jericho: A Cautionary Tale', *Strata*, 27 (2009)

Engelberg-Baram, Orit. 'Parashat Sdom: tmorut ba-yachas ha-zioni la-sviva derech mikreh mivchan meizamay pitoach shel yam-hamelach be-sdom [The Sodom Affair: Changes in the Zionist Approach to the Environment through the Case Study of Development Initiatives of the Dead Sea at Sodom]', *Muza*, 5 (2022)

Gavish, Dov. *Melah Haaretz* [Salt of the Land] (Jerusalem, 1995)

Gavish, Dov and Ada Schein. *Sipuro shel mifal ha-tamrukim le-hof yam ha-melach* [The Story of the Cosmetics Factory on the Shore of the Dead Sea] (Jerusalem: Ariel, 2007)

Gil-Har, Yitzhak. 'Boundaries Delimitation: Palestine and Trans-Jordan', *Middle Eastern Studies*, 36:1 (2000)

Giladi, Yair. *Haim hadashim le-yam ha-melah* [New Life for the Dead Sea] (Jerusalem, 1998)

Haddadin, Munther. 'Water Issues in Hashemite Jordan', *Arab Studies Quarterly*, 22 (2000)

Al Husseini, Jalal. 'The Dilemmas of Local Development and Palestine Refugee Integration in Jordan: UNRWA and the Arab Development Society in Jericho (1950–80)', *Jerusalem Quarterly*, 93 (2023)

SELECT BIBLIOGRAPHY

Ionides, M.G. 'The Disputed Waters of the Jordan', *Middle East Journal*, 17:2 (1953)
Jaber, Atwa. 'No Bridge Will Take You Home: The Jordan Valley Exodus Remembered through the UNRWA Archives', *Jerusalem Quarterly*, 93 (2023)
Kedar, B.Z. *Mabat ve'od mabat al eretz yisrael* [Looking Twice at the Land of Israel] (Jerusalem, 1991)
Koren, David. *Mol ha-yam ha-kaved: hevrat ha-ashlag ve-ha-'Haganah'* [Facing the Heavy Sea: The Potash Company and the 'Haganah'] (Tel Aviv, 1991)
McCarthy, Justin. *The Population of Palestine* (New York: Columbia University Press, 1990)
Ne'eman, Yuval and Israel Schul. 'Israel's Dead Sea Project', *Annual Review of Energy* (1983)
Norris, Jacob. 'Toxic Waters: Ibrahim Hazboun and the Struggle for a Dead Sea Concession, 1913–1948', *Jerusalem Quarterly* (2011)
Novomeysky, Moshe. *Given to Salt: The Struggle for the Dead Sea Concession* (London, 1958)
Oren, Ahron and Antonio Ventosa. 'Benjamin Elazari Volcani (1915–1999): Sixty-three Years of Studies of the Microbiology of the Dead Sea', *International Microbiology*, 2 (1999)
Rabineau, Shay. *Walking the Land: A History of Israeli Hiking Trails* (Bloomington: Indiana University Press, 2023)
Reguer, Sara. 'Rutenberg and the Jordan River: A Revolution in Hydro-Electricity', *Middle Eastern Studies*, 31:4 (1995)
Schein, Ada and Dov Gavish. 'Briut ve-yofi ke-manof le-pitoah hevel yam ha-melah [Health and Beauty as an Impetus to the Development of the Dead Sea Region]', *Karka*, 66 (2009)
Shapiro, J.M., Z. Saliternik and S. Belfekman. 'Malaria Survey of the Dead Sea Area during 1942, including the Description of a Mosquito Flight Test and its Results', *Transactions of the Royal Society of Tropical Medicine and Hygiene*, 38:2 (1944)
Shehadeh, Raja. *A Rift in Time: Travels with My Ottoman Uncle* (London: Profile Books, 2010)
Sosland, Jeffrey K. *Cooperating Rivals: The Riparian Politics of the Jordan River Basin* (State University of New York Press, 2007)
Tal, Hila. *Mahapecha Be-Sdom* [Renaissance at Sodom] (Ben Gurion Research Institute, 2010)
Volcani, Benjamin Elazari. 'Algae in the Bed of the Dead Sea', *Nature*, 22 June 1940
Wilkansky, Benjamin. 'Life in the Dead Sea', *Nature*, 12 September 1936
Willner, Samuel E. 'Maritime Navigation, Energy Security, and Peace Dividend: Exploring the Historic Developments of the Dead Sea Conveyance Project', *Journal of the Middle East and Africa*, 8:1 (2017)

CONTEMPORARY ENVIRONMENTAL DEGRADATION AND RESTORATION EFFORTS

Allan, John Anthony, Abdallah I. Husein Malkawi and Yacov Tsur. 'Red Sea–Dead Sea Water Conveyance Study Program: Study of Alternatives – Final Report', March 2014
Ashford, J. L. and B. A. Boocock. 'Arab Potash Solar Evaporation System: Construction', *Proceedings of the Institution of Civil Engineers* (1984)
Closson, Damien et al. 'Subsidence and Sinkhole Hazards in the Dead Sea Area, Jordan', *Pure and Applied Geophysics*, 162 (2005)
Courcier, Rémy, Jean-Philippe Venot and François Molle. 'Lower Jordan River Basin (in Jordan): Changes in Water Use and Projections (1950–2025)', International Water Management Institute, report no. 9 (2005)

SELECT BIBLIOGRAPHY

Ecopeace. 'Ecopeace's Vision for Stabilizing the Water Level of the Dead Sea', 2018

Haddad, Naif A., Sharaf A. al-Kheder and Leen A. Fakhoury. 'Al Mujib Natural Reserve in Jordan: Towards an Assessment for Sustainable Ecotourism Management Plan Utilizing Spatial Documentation', *Natural Resources and Conservation*, 1:3 (2013)

Haddadin, Munther J. 'A Jordanian Socio-Legal Perspective on Water Management in the Jordan River–Dead Sea Basin', in Clive Lipchin, Deborah Sandler and Emily Cushman (eds), *The Jordan River and Dead Sea Basin: Cooperation amid Conflict* (Dordrecht: Springer, 2009)

Khlaifat, Abdelaziz et al. 'Mixing of Dead Sea and Red Sea Waters and Changes in their Physical Properties', *Heliyon*, 6:11 (2020)

Ministry of Environmental Protection (Israel). 'Joining Forces to Address Climate Change Impact: Israel and Jordan to Cooperate on Jordan River Restoration', 17 November 2002

Oroud, Ibrahim M. 'The Future Fate of the Dead Sea: Total Disappearance or a Dwarfed Hypersaline Hot Lake?', *Journal of Hydrology*, 623 (2023)

Raz, Eli. 'The Future of the Dead Sea: Is the Red Sea–Dead Sea Conduit the Right Solution?', in Clive Lipchin, Deborah Sandler and Emily Cushman (eds), *The Jordan River and Dead Sea Basin: Cooperation amid Conflict* (Dordrecht: Springer, 2009)

Royal HaskoningDHV. 'Red Sea Dead Sea Water Conveyance Study', submitted to the European Investment Bank, 6 December 2017

Shalev, Eyal (ed.). 'Tgovat ha-tashtit be-agan yam ha-melach le-shinoyim ha-tive'yim ve-maase yedei adam [The Response of Infrastructure in the Dead Sea Basin to Natural and Man-Made Changes]', *Geological Survey of Israel*, January 2022

Taqieddin, Salah, Nabil S. Abderahman and Mohammad Atallah. 'Sinkhole Hazards along the Eastern Dead Sea Shoreline Area, Jordan: A Geological and Geotechnical Consideration', *Environmental Earth Sciences*, 39:11 (2000)

INDEX

Abbasid Caliphate 5, 110–13, 237
Abdullah I, King 169, 171, 179–80, 188
Abdullah II, King 217
Abdulhamid II, Sultan 157–8
Abraham 42
Acre 104–5, 115
Admah 43
Aelia Capitolina 93
Asfaw, Empress Menen 191
Agnon, Amotz 56
Agrippa, Marcus 78
Ahava 214
Ain es-Sultan 19, 23, 26, 32, 51–2, 119
 see also Elisha's Spring
Ain Jidy see Ein Gedi
Albright, William F. 56
Alexander Jannaeus 76, 236
Alexander the Great 6, 70–1, 73 236
Allen, William 142, 146–7
Allenby, Adelaide Mabel (Adela) 165
Allenby Bridge 158, 162, 176, 192
Allenby, General Edmund 165–6
Almog, Kibbutz 225–6
Alon, Azaria 210
Alon, Yigal 178
Amaziah, King 53, 236
Amman 24, 36, 67, 74, 99, 104, 157, 162, 166, 180–1, 192–3, 219, 232
Amos (biblical prophet) 55–6, 65
Antigonus (Hasmonean ruler) 77–8
Antigonus Monophthalmus (Antigonus the One–Eyed) 71

Antiochus IV Epiphanes, King 75
Antiochus VII, King 76
Antipas, Herod 58
Antipater 77
Antony, Mark 77–8, 80, 86
Aqaba 11, 104, 133, 145, 147, 158, 162, 202, 219–20
Aqbat Jabr 162, 180, 190, 225–6
Arab League 196–7
Arab Legion 177–8
Arab Potash Company 202–3, 205–6, 220, 229, 238
Arabah (Arava), Valley 8, 12, 24, 33, 104, 133, 143–5, 147, 178
Arabia 86, 103, 107, 109, 112, 117, 143, 166, 171
 Roman province of 95
 Arabian leopard 112
Arabian Plate 11–12
Aral Sea 200
Arava Institute for Environmental Studies 87–8 232
Ariha see Jericho
Aristotle 2, 84, 137
Arnon see Mujib, Wadi
asphalt 69–72, 78, 84, 100, 118, 127, 129, 138, 149, 159, 211
 Lake Asphaltites 6, 69, 100
Atlantic Ocean 3, 13, 15, 20
Augustus (Octavian) 78
Auja 225
Australia 165–6

INDEX

Ayalon, Merav 206
Ayyub Dynasty 114, 122

Bab edh-Dhra 24, 33–35, 45, 203
Babatha 94–95, 117
Babylon 63, 128
  Babylonian Empire 57, 60, 70, 236
Baghdad 60, 110, 113, 122
Bajnotti, Paolo 135
Baldwin I, King 116
Baldwin IV, King 121
balsam 85–7, 93, 100, 106, 110–11, 227
Bani Hamida tribe 53–4
Bar Adon, Pesach 31
Bar Kokhba 92–3, 95–6, 102, 237
Bar-Oz, Guy 108
Bartov, Yuval 16
Bar-Yosef, Ofer 17, 22
Baybars, Sultan 49, 121–3, 237
Bedouins 53–54, 60–61, 63, 72, 89–91, 117, 123–5, 129, 132, 150–155, 157–8, 168, 179, 210–11
Beek, William G. 133–4
Beirut 133–4
Beisan Valley 13
Beit HaArava, Kibbutz 171, 176, 178
Belfer-Cohen, Anna 18
Ben Gurion, David 174, 177, 186
Ben-Yair, Eleazar 81, 83
Benjamin of Tudela, Rabbi 124–5
Berlin 62–3, 232
de Bertou, Jules 133–4
Bethany Beyond the Jordan 36–38, 97, 162, 216–17, 239
Bethlehem 63, 72, 91, 104, 106, 127, 151, 169
Birger, Hillel 209
bitumen *see* asphalt
Blake, George Stanfield 175
Blanckenhorn, Max 160
Blass, Simcha 183, 186, 194
Blum, Yehuda Zvi 196–7
Bourcart, Abraham Max 148
British Empire 135, 146–7, 160, 163–9, 171, 173, 175, 238
  British Mandate (Palestine) 169–71, 175–7, 179, 182, 190, 209, 238
British Library 61
British Museum 72
Brock, Major Reginald Walter 167–8
bromine 160, 167–8, 171, 173, 202
Burchard of Mount Sion 120, 124–5

Byzantine Empire 5–6, 38, 43, 48, 70, 96–103, 106–9, 111–13, 120, 163, 203, 226, 237

Caesar, Julius 77
Cairo 110, 113, 123, 169
  Cairo Genizah 249 n. 44
Callirrhoe 36, 79, 100–1, 114, 212, 226
Canaan 5, 23, 41, 50
Carnallite 167
Carne, John 156
Chalcolithic 26–32, 235
China 88, 168
Choziba, Monastery of 97
Christianity 6, 38–9, 44, 58–61, 76, 93, 96, 101–3, 111–12, 117, 119, 156–7, 189–91, 208, 216–17
Churchill, Winston 169
cities of the plain 41–2, 44–5, 47, 105, 127–8, 149
Clapp, Frederick Gardner 44
Cleopatra, Queen 78, 80, 86, 213, 236
Clermont-Ganneau, Charles 53–4, 61, 163
climate change
  contemporary 89, 217–18
  historic 4–5, 14–15, 17, 20, 25, 29, 32, 41, 98–9, 107–8, 123, 235, 242 n. 22
Conder, Claude R. 44, 61–2, 135
Constantine I, Emperor 96, 237
Costigan, Christopher 139–40
Crusaders 1, 5, 12, 44, 59, 113, 115–23, 128, 149–50, 237
Cypros (Herod's mother) 77, 79
Cyprus 72
Cyrus the Great 70, 236

Daliyeh, Wadi 73
Dalman, Gustaf 65
Damascus 36–7, 49, 65, 104, 110, 113, 121, 166
Damascus Document 65–6, 249 n. 44
Damascus Gate (Jerusalem) 134
Dayan, Moshe 224
date palms 1, 57, 69, 87–9, 115, 206, 225, 227, 232
David, King 39, 52–3, 57–8
  City of David 56
Dead Sea Scrolls 1–2, 25, 59–67, 82–3, 89–91, 236 249 n. 42
Dead Sea Works 162, 200–3, 220, 228, 238
Decapolis 107

INDEX

Deeb, Shukri 170, 174
Degania Dam 162, 183, 187, 204, 221–2
Deir Alla 36, 56
Dershowitz, Idan 62
Desalination 146, 219, 221–2
Deuteronomy 48, 61–2
edh-Dhib, Muhammed 63
Dibon (Dhiban) 36, 53–5
Dio, Cassius 92–3
Diocletian 96
Diodorus Siculus 71–2, 85, 129
Divine Comedy 77
Dok (Duk/Dagon) 76–7, 236
Drabsch, Bernadette 27
Drake, Charles F. Tyrwhitt 136, 151, 154
drought 25, 34, 41, 107
Dunaliella 141, 204–5

Early Bronze Age 29–30, 32–34, 39, 45, 51, 203, 235
earthquakes 4, 12, 34, 46, 51, 55–6, 78, 82, 112–13, 142, 236–8
East Ghor Canal (King Abdullah Canal) 162, 188–92, 194, 238
EcoPeace 220, 229
Edom 5, 36, 40, 52, 55, 69, 77, 149
Edwards, Phillip 19
Egypt 32–35, 40–1, 48, 61, 63, 69, 71–75, 78, 85–6, 110–11, 113, 115–16, 121–2, 129, 160, 166, 188, 190, 195, 236
Ein Bokek 162, 195, 213, 227–8
Ein el-Sultan (refugee camp) 180
Ein Gedi (Ain Jidy) 8, 14, 29, 31, 35–6, 39, 56–7, 70, 81, 85–7, 91, 93–5, 101–4, 110, 128, 162, 178–9, 181, 205
  Ein Gedi oasis 87, 99, 123, 179, 194, 206, 209–10
  Kibbutz Ein Gedi 57, 69, 179, 194, 204, 206, 212–14, 226–7
  Ein Gedi hot springs 212, 227
Ein Feshkha 8, 35, 56, 97, 104, 131, 162, 164, 172, 209–10
  Ras Feshkha 44, 209
Eisenhower, Dwight 188
Elijah (biblical prophet) 39, 49, 79
Elisha (biblical prophet) 37, 39, 49–50, 52
  Elisha's Spring 52, 79, 224
Ellenblum, Ronnie 12, 233
Emery, Kenneth Orris 109, 184, 204
Engelberg-Baram, Orit 199

Epiphanius of Salamis 60
Ethiopia 85, 117, 191, 217
Eusebius 50, 86, 93, 98, 100, 102
Ezekiel (biblical prophet) 57, 100

Fabri, Felix 116, 125–6, 130
Fatah 186
Fatimid Caliphate 113, 115–18
Faynan, Wadi 8, 24, 31
Finkelstein, Israel 52
Freund, Raphael 11
First World War 6, 65, 160, 163–9, 172, 182, 238
Flaubert, Gustave 150
Forder, Archibald 155, 158
Frankish Kingdom of Jerusalem 113–14, 121, 237
  see also Crusades
France 74, 133, 135, 140, 147, 165
Fulcher of Chartres 116–17, 120
Fulk, King 117
Fuller, Thomas 44, 128

Gabriel Inscription 249 n. 42
Galen 86, 114, 126
Galilee (region) 18, 77, 80
Galilee, Sea of 8, 12, 15–16, 36, 104, 134, 139, 159, 182–3, 186–7, 219, 221–2
Garfinkel, Yosef 27–8
Garstang, John 20, 51
Gayer, Asaf 95
Gaza 162, 195–6, 223, 225
Germany 61, 167, 173
Gesher Benot Ya'aqov 8, 17, 27
Gilad, Moshe 207–8, 227, 232
Gilgal 38, 49
Ghassulian culture 26–27, 30–31, 235
Ghawarna 151, 154
Ghor *see* Jordan Valley
Ghor al-Haditha 162, 193, 206
Ghor es-Safi 2, 8, 24–5, 43–4, 104, 116, 152, 156, 162, 171–2, 174–5, 178, 203, 216, 226
  see also Zoar
Glubb, General John Bagot 177
Göbekli Tepe 23
Golb, Norman 66
Goldenberg, Leehee 220, 222, 232
Great Salt Lake (Utah) 200
Greek-Orthodox Christianity 38, 59, 99, 126–7, 216
Guttman, Shmarya 178–9

281

INDEX

Hadas, Gideon 69, 87, 214–15, 231
Hadrian 92–3
Haganah 175
Haidar, Prince Abdul Majid 189
Haifa 147, 185
HALO 216–17, 232
Hammeh, Wadi 8, 18–19
al-Hasa, Wadi 8, 25, 102, 104, 108, 162, 193
Hashem, Ibrahim 178
Hashemites 166, 169
  see also Jordan, Hashemite Kingdom of
Hasmoneans 5, 74–78, 80, 118, 236
Hasselquist, Fredrik 137–8, 156
Hasson, Nir 225–6
Hayes, James B. 262 n. 43
Hazboun, Ibrahim 169–70, 173
Hazeva 8, 15
Hazor 36, 41, 56
Hebrew Bible 1–2, 5–6, 20, 37–45, 47–53, 55–7, 59–62, 64–5, 74, 93, 97, 128
Hebron 104, 133, 151, 162, 171, 181
Hellenistic period 39, 67, 69–74, 101, 135
Heraclius, Emperor 103, 109, 237
Hermon, Mount 104, 137
Herod, King 77–80, 82, 86, 212, 236
Herodotus 72
Herzl, Theodor 148, 159, 185, 194–5, 238
Hieronymus of Cardia 71
Hill, John Gray 135–6
Hirschfeld, Yizhar 76, 99
Hisban 18, 104, 122
Hisham's Palace 104, 111–12, 224
Holocaust 184
Holocene 4, 14, 20, 27, 29
Horowitz, Aharon 12, 15–16, 144
hot springs 79, 100, 212–13, 226–7
Hudleston, Wilfrid H. 143, 147
Hull, Edward 143–5
Hunt, William Holman 149–50
Huntington, Ellsworth 9, 145, 154
Hussein, King 188–9, 192, 215
Husseini, Hajj Amin al- 175
hydroelectricity 148, 182–84, 191, 194–95

ibex 210–11, 227
Ibn Hammad Dam 193
Ibn Hawqal 114–15

al-Idrisi, Muhammad 115, 118
Idumaeans 1, 77
India 114–15, 135, 146–7, 166
Indian Ocean 3, 11
Inter–seas Water Conduit 195–97
Ionides, Michael G. 188
Iraq 89, 112, 115, 175
Iraq al–Amir 74
Irbid 18
Iron Age 35, 40–1, 44, 56–7
Islam 6, 39, 43–4, 48–50, 105, 109–15, 119, 121, 123, 208
Israel (biblical kingdom) 39, 41, 52–6, 64
  Israelites/Children of Israel 37–8, 40, 48–51, 54–5, 64, 128
Israel, State of 2, 5, 18, 56, 67, 87, 90–1, 162, 173–4, 177–81, 186–92, 194–97, 199–203, 207–10, 213–26, 228, 238–9
Israel Antiquities Authority 75, 223
Israel Chemicals Ltd (ICL) 202, 220, 228–9, 264 n. 6
Israel Defense Forces (IDF) 91, 178–9, 181, 190–1, 200, 216, 225
Israel Museum 31, 67
Israeli Water Authority 182, 202, 221, 264 n. 1
Istakhri 119
Italy 92, 106, 135, 175, 189

al-Jadida 165
Jaffa 38, 104, 134, 158, 174
Jahalin tribe 151–2
James, Sir Henry 134
Japan 224
Jawhariyyeh, Wasif 165
Jehoshaphat, King 55
Jeremiah (biblical prophet) 60, 66, 206
Jericho 2, 8, 19–26, 28, 32–9, 47–52, 58–60, 62–3, 70, 73–4, 76, 78–9, 85–8, 93, 97–8, 100–4, 110–12, 114–16, 118–20, 122–3, 126, 128, 138, 142, 148, 151, 153–8, 162–6, 168, 170–2, 176–7, 224–5
  Jericho Conference (1948) 179
  planned airport 180–1
  refugee camps near 180, 190
  Ruha 23, 35
Jerusalem 8, 20, 24, 36, 38, 49, 53, 56–7, 59, 61, 63, 66–7, 72, 74–6, 78–81, 88, 90, 93, 97, 100, 104–5, 108, 110,

INDEX

113–15, 117–21, 124–7, 129, 131, 133–4, 140–2, 151, 153, 157–8, 162, 165, 170–1, 175, 177–8, 180, 224–5, 236–7
East Jerusalem (Jordanian) 67, 89, 180
Jesus 37, 58–9
Jewish-Roman War 80–1, 93
Jezreel Valley 8, 13, 104, 143, 147–8
John the Baptist 37, 58, 150
   religious sites commemorating 38, 216–17
Johnston, Eric 188–9
Joram, King 55
Jordan, Hashemite Kingdom of 2, 5, 24, 43, 56, 74, 91, 179–81, 188–93, 196, 199, 202–3, 206, 208, 211–12, 214–16, 218–22, 226, 229, 232
see also Transjordan
Jordan Museum 67
Jordan River 2, 11, 18, 26, 28–9, 34, 37–9, 48–9, 64, 75, 100, 112, 116, 119–21, 125, 127–8, 132–4, 137–9, 141, 152, 154, 158–9, 164, 166, 168, 172, 177, 182–3, 185–90, 192, 194, 200, 208, 215–16, 218, 220–2
Jordan Valley 4, 11, 13–14, 16, 18–20, 23, 27–9, 46, 48, 51, 56, 109, 112, 115–16, 119, 135, 142–5, 147–8, 151, 165–6, 171, 180, 182, 184, 186–91, 193–4, 222, 225, 232, 238
Josephus (Yosef ben Matityahu/Flavius Josephus) 42–3, 57–8, 65, 76–7, 79–87, 89, 92, 129, 211
Joshua 20, 38–9, 49–51, 170
Judea 5, 75, 77, 81, 83, 85–6, 92, 150
Judean Desert 15, 75, 91–2, 103, 108, 209
Judah 5, 41, 53, 55, 60, 236
Judaism 1, 6, 39, 44, 59–60, 64–5, 73–5, 77, 80, 83, 85–6, 90, 92–3, 95, 101–3, 112, 127, 148–9, 163, 224
Justinian I, Emperor 103, 106, 237

Kabátník, Martin 256 n. 56
Kalia hotel 173, 213
Kalia, Kibbutz 165, 227
Kaptijn, Eva 15, 30
Karak (Kerak) 96, 98, 104, 121–2, 151, 155, 158, 165, 211, 217
Kenyon, Kathleen 20–2, 224
Khirbet Ataruz 36, 55

Khirbet al-Mafjar *see* Hisham's Palace
Khirbet en-Nitla 112
Khirbet Mazin 36, 76, 99, 118, 162, 206
Khirbet Qazone 102
King Talal Dam 192–93
Kitchener, Horatio Herbert 135
Klein, Eitan 75
Klein, Frederick Augustus 53

Lake Amora 14
Lake Chad 200
Lake Samra 14, 235
Lake Urmia 200
Landau, Sigalit 227
Langgut, Dafna 27–8
Langotzky, Moshe (Musia) 174
Lansky, Nadav 207–8
Lapp, Paul 33
Lartet, Louis 14, 143
Late Bronze Age 32, 35, 39–41, 47, 236
Le Roux, Magdel 41
Leal, Beatrice 100
Lebanon 11, 34, 72, 143, 182, 188, 192
Lewinsky, Elhanan Leib 159
Linnaeus, Carl 137
Lisan, Lake 14–17, 143, 221, 235
Lisan Peninsula 8, 14, 16, 24, 33, 36, 100, 102, 104, 107, 136, 153, 158, 162, 203, 205–6, 219, 249 n. 42
Litt, Thomas 29
Lortet, Louis 140
Lot (biblical figure) 1, 6, 41–4, 85, 128
   Sanctuary of Agios Lot 43, 97, 113, 216, 226
Louvre, Museum 55, 63
Lowdermilk, Walter Clay 184–85, 194, 262 n. 43
Lynch, William Francis 134–5, 151
Lynch Strait 104, 107, 123, 134, 162, 202–3, 206, 208

Maccabees *see* Hasmonaen
Machaerus 36, 58, 77, 79, 81, 101, 150
Madaba 36, 54, 99, 157, 162, 211, 216
   Madaba map 99–101, 103, 105, 231
Madden, Richard Robert 138
al-Maghtas *see* Bethany Beyond the Jordan
Magness, Jodi 66
Mahoza 94
Ma'in hot springs (Hammamat Ma'in) 212

INDEX

el-Makkukh, Wadi 28, 88
Maklef, Mordechai 200–1
malaria 166, 172
Malmesbury, William of 117
mappa mundi 127–8
Maqarin Dam 193
Masada 11, 36, 77–83, 86–8, 91, 97, 111, 162, 176, 178, 202, 209, 210, 237
Masterman, Dr Ernest William Gurney 131, 137, 145, 158
Matari, Yusuf 218
Matariyya 110–11
Maundrell, Henry 129, 137–8
Mazraa 104, 158
Mecca 49, 121
Medina 112, 121
Mediterranean Sea 3, 11–13, 17–18, 21, 27, 29, 38, 40–1, 48, 106, 108–9, 114, 119–20, 122–3, 132–4, 139, 143, 147–8, 158, 168, 184, 185, 194–7, 220–2
Megiddo 36, 41, 56, 166
Mehmed V, Sultan 159
Melisende, Queen 117
Mesha, King 54–5, 236
  see also Moabite Stone
Mesopotamia 32, 63, 109
Methuselah 87–8
Meyhar, Munqeth 218
Middle Bronze Age 24, 32, 35, 41, 46–7, 51, 235
Milankovitch Cycles 17, 242 n. 22
Milhat Sdom 229
Mishnah 59
Mithen, Steven 23
Mitzpe Shalem 162, 214
Moab 1, 5, 36, 40, 48, 53–5, 61–2
Moabite Stone 53–5, 61–2, 152, 236
Mongols 122–3, 237
Moore, George Henry 133–4
Moshiov, David 213
Moses 48–50, 122, 128, 170
  see also Nabi Musa
Mu'awiya I 110
Muhammad 110, 112, 237
Mujib, Wadi (Arnon) 8, 36, 54, 61, 98, 158, 162, 193, 211, 226
Mujib Dam 162, 193
al-Muqaddasi 105, 115–17, 126
Murabba'at, Wadi (Nahal Darga) 36, 75, 90–3, 123, 207
Museum at the Lowest Place on Earth 226

Naaman 37, 39
Naaran 101–3, 163
Nabataeans 1, 69–71, 75–8, 95, 102
Nabi Musa 36, 48–9, 104, 122–3, 162, 165, 237
Nadaud, Clifford E. 38
Nahal Hever 36, 91, 93, 95
Nahal Mishmar 8, 31
Naharayim 162, 183, 191–2
Nahash the Ammonite 64
National Water Carrier (Israel) 186–7, 194, 200, 222, 238
Natufians 17–19, 235
Nasser, Gamal Abdel 188
Nebo, Mount 36, 48, 54, 97, 104, 112, 128
Ne'eman, Yuval 195
Neev, David 109, 184, 204
Negev Desert 36, 96, 99, 104, 108, 162, 174, 178, 186, 194, 223
Neolithic 6, 8, 20–3, 26, 203, 224, 243 n. 33
Neot HaKikar 228
Neumann, Frank 29
New Testament 43, 58–9
New Zealand 165
Nigro, Lorenzo 21, 23, 33–4, 47, 51
Nissenbaum, Arie 72
North Africa 17, 106, 109, 113, 143
North America 6–7, 20, 25, 82, 109, 132, 150, 154, 235
Novomeysky, Moshe 160, 169–75, 177, 199–200, 213, 228
Numeira 8, 24, 34–6, 45

oil 164–5, 174–5, 195, 197
olives 27–9, 82, 95, 99
Omri, King 54–5
Origen of Alexandria 60
Oron, Asaf 118, 215
Oroud, Ibrahim 217–18
Oslo Accords 215, 224
Ottoman Empire 5, 14, 38, 49, 54, 124, 129, 135, 147–50, 152–4, 157–60, 163–6, 237
Oultrejordain 5, 121

Palestine 5, 14, 44, 51, 90, 92, 96, 99, 101, 107, 109, 112–14, 117, 119, 135, 138–9, 143, 148, 156, 159–60, 165–6, 169–77, 179, 182–5, 209, 238
  Palestinian Authority 51, 215, 224–5
  Survey of Western 44, 51, 135–6, 143, 151, 154

284

INDEX

Palestine Archaeological Museum (Rockefeller Museum) 67, 72, 89
Palestine Electric Corporation 183
Palestine Exploration Fund (PEF) 20, 44, 51, 53, 61, 131, 134–7, 143, 151, 154, 232, 238
PEF Rock 131, 135, 137, 238
Palestine Liberation Organization (PLO) 191–2, 215
Palestine Potash Company 162, 170–4, 176–7, 179, 184, 200, 238
Palestinian refugees 180, 187, 189, 190
Palmer, Edward Henry 136
Parthians 77, 85, 236
Pausanias 9
Perea 5, 75
Pergamonmuseum (Vorderasiatisches Museum) 63, 232
Persian Achaemenid Empire 70, 72–4, 101, 236
Persian Gulf 11, 124
Persian–Sasanian Empire 108–9, 237
Petermann, Julius Heinrich 53
Petra 24, 36, 70, 102
pillar of salt (Lot's wife) 1, 42–3, 125, 125, 127, 218
plague 106–7, 123
plate tectonics 10–11, 13, 16, 141–3, 145, 195
Pleistocene 14–15, 205, 221
Pliny the Elder 65, 79, 84–7, 129
Pococke, Richard 129–30
Politis, Konstantinos D. 44–5, 113, 116
potash 159, 167–8, 170, 172–3, 175, 177, 195, 201–3, 238
see also Arab Potash Company and Palestine Potash Company
Pre-Pottery Neolithic
Pre-Pottery Neolithic A (PPNA) 21–2, 24, 235
Pre-Pottery Neolithic B (PPNB) 21–2, 24–5, 235
Proskynetaria 126–7, 256 n. 56
Prussia 53–4, 61
Ptolemaic Egypt 73–4, 77–8
Ptolemy II Philadelphus 73, 236
Ptolemy VI 75, 236
Ptolemy son of Abubos 76–7

al-Qari, Salim 53–4, 61
Qasr al-Abd 36, 74
Qasr al-Yahud 36–8, 162, 190, 216, 232

Qasr Bshir 36, 98
Qelt, Wadi 36, 76, 79, 97, 104, 111
Queen of Sheba 85
Quennell, Albert 11
Qumran 36, 62–7, 76–7, 83, 86–90, 97, 128, 165, 236, 249 n. 44
Quran 43
Quruntul *see* Temptation, Mount of

Rabin, Yitzhak 215
Rahav, Oded 218, 232
Rainey, Anson F. 40
Rast, Walter 33, 45
Raz, Eli 204–5, 222, 232
Red Sea 11, 98, 108, 121, 133, 142–3, 145, 147, 195–6, 202, 219–20
Red Sea–Dead Sea Water Conveyance 219–20
refuge caves 1, 92–3, 95
refugee camps 180, 190, 225
Reynald of Châtillon 121
Robinson, Edward 50, 82, 156
Roman Empire 1, 5, 9, 39, 50, 57–8, 60, 67, 70, 77–96, 98–9, 101, 107, 109, 114, 126, 135, 236
   Western Roman Empire 109
Route 65 226
Route 90 15, 76, 131, 206, 227–8
Royal Society for the Conservation of Nature (Jordan) 211
Rubin, Rehav 98
Ruha see Jericho
Rujum el-Bahr 36, 101, 104, 128, 135–6
Russell, Irina 41
Russell, Sir John 173
Russia 135, 157, 182
Russian Orthodox Church 59
Rutenberg, Pinhas 182–3, 194

Saladin (Salah al-Din) 114, 121, 237
Sallon, Sarah 88–9, 232
salt 1–2, 4, 13–14, 16, 29, 46–7, 50, 57, 70, 94, 100, 114, 118, 120, 129, 139–41, 144–6, 149, 158–60, 167–8, 170, 185, 204–5, 208, 211–12, 218, 221, 227–8
   City of 159
   Salt Sea 5, 42, 58
   Valley of 52–3
As-Salt 104, 122, 154, 162, 171
Salti, Lieutenant Muwaffaq 181, 215

INDEX

Samaritans 73, 101, 249 n. 44
Samuel, Book of 52, 57, 64
Saracens 98, 103, 108, 117
Saul, King 57, 64
de Saulcy, Louis Félicien 149, 152
Schaub, Thomas 33, 45
Scott, Walter 149
Sebbeh *see* Masada
Second World War 173, 175–6, 184
Seetzen, Ulrich Jasper 82, 138, 151
Segor *see* Zoar
Selassie, Haile 191
Seleucid Empire 74–7
Seljuk Empire 113
Sellin, Ernst 20
Shahin, Khalil Iskander 'Kando' 72, 89, 91
Shahwan, Husni 224
Shalom al Israel synagogue 102, 112, 224
Shapira, Moses 61–2
Shertok (Sharett), Moshe 174
Sheshonq I (Shishak), Pharoah 41, 236
Shivta 104, 108
Siberia 46, 160, 169
Sicarii 81
Siddim, Valley of 42–3, 149
Silva, General Lucius Flavius 81
Silvia, Phillip 45, 231
Simon Thassi 76, 236
Sinai Desert 103, 195
Sinai Subplate 11
sinkholes 205–8, 212
Sivan, Hagith 97
Six-Day War 67, 91, 187, 190, 196, 210, 214, 216, 224, 238
slavery 37, 73–4, 81, 116
Society for the Protection of Nature in Israel 210
Sodom, Mount (Jebel Usdum) 8, 13, 16, 104, 114, 120, 133, 149, 162, 167, 170, 172, 175–6, 178, 184, 200, 202, 205, 228
Sodom and Gomorrah 41–7, 50, 59, 84, 105, 127, 149, 156, 208–9
  apple of Sodom (*Calotropis procera*) 84–5, 227
Solomon, King 57, 85
Solowey, Elaine 88, 232
southern Levant 2, 14–15, 17, 26–7, 30, 55, 70–4, 79, 89, 106–9, 113–15, 121, 123–4, 134, 146, 150–1, 157, 169, 180, 214, 237

Standard Oil 164
Stein, Mordechai 15
Strabo of Amasia 71, 84–5
Suez Canal 135, 146–7, 238
sugarcane 1, 106, 114–16, 123
Sukenik, Eliezer 63, 90
Suleiman the Magnificent 124, 237
sulphur 1, 43, 127, 149, 209
Sweimeh (Suwayma) 215, 226–7
Syria 34, 37, 65, 92, 110, 114–15, 117, 119, 121–2, 143, 160, 171, 182, 186–8, 190, 193, 195

Ta'amireh tribe 72, 89–90, 151
al-Taji, Abdul Rahman 174
Tall el-Hammam 8, 24, 34–6, 45–7, 236
Talmud 85
al-Tamimi, Muhammad ibn Sa'id 114, 119
Tannur Dam 162, 193
Taylor, Joan 128
Tekoa (Thecua) 36, 55, 94, 104, 117–18
Tel Tsaf 8, 28
Tell es–Sultan 8, 19–21, 23, 26, 32, 47, 50–2, 59, 76, 102, 112, 180, 224
Templars 120–1
Temptation, Mount of 23, 58–9, 76, 97, 120–1, 224
Tennessee Valley Authority 184, 262 n. 43
Tethys Ocean 10–11, 13
Thucydides 81
Tiberius, Emperor 58
Timothy I of Baghdad 60
Titus 80, 86
Tobiad family 74
Trajan, Emperor 95
Transjordan 5, 14, 40, 54, 96, 99, 143, 158, 165–6, 169
  Emirate of 169, 171–2, 174–5, 177–9, 182–3, 187–8
Tristram, Henry Baker 58, 138–9, 142–3, 152–3, 155–7, 228
Tuleilat Ghassul 8, 24, 26–8, 30–2
Tulloch, Major Thomas Gregory 169–70, 172
Tunguska 46
Tunick, Spencer 218
Turkey 22–4
  see also Ottoman Empire
Twain, Mark 50, 153
Tyr 125–6
Tyros 74

286

INDEX

Umayyad Caliphate 5, 48, 110–12, 224
United Nations (UN) 176, 192, 196–7
  UN Partition Plan (1947) 176, 178–9, 238
United Nations Educational, Scientific and Cultural Organization (UNESCO) 51–2, 216, 224
United Nations Relief and Works Agency (UNRWA) 180
United Nations Special Committee on Palestine (UNSCOP) 176
United States 38, 50, 67, 90, 134, 142, 168, 184–6, 188–9, 200, 208, 223
Usdum, Jebel *see* Mount Sodom
Ussishkin, David 31
Uziel, Joe 56, 223
Uzziah, King 55–6

Warren, Charles 20, 51, 53–4
Watzinger, Carl 20
Wegener, Alfred 141, 145
Weizmann, Chaim 185
West Bank 19, 48, 73, 162, 178–9, 181, 186–7, 190, 214, 216, 225
Wilde, Oscar 58, 150
Wilhelm II, Kaiser 158
William of Tyre 120
Wilson, John 85, 151, 156–7
World Bank 201, 219

Vainstub, Daniel 66
de Vaux, Roland 63, 66, 89–90, 93
de Vitry, Jacques 105, 116, 120, 126
Vernon, Roland V. 170

Vespasian 80, 84, 86
Volcani (Wilkansky), Benjamin Elazari 140–1, 204
Vörös, Győző 79

Yadin, Yigael 82–3, 88, 90–3
yahad 65–6, 83, 249 n. 42
Yaqut al-Rumi al-Hamawi 116
Yarmouk River 162, 164, 183–4, 187–9, 191–4, 204
Yehud 70
Yitav 225
Younger Dryas 15, 20, 235

Zacchaeus 59, 156, 224
Zacharias of Sakha 110
Zahrat adh-Dhra 8, 24
Zanclean Flood 13
Zara 104, 114, 162, 212, 215, 226
  see also Callirrhoe
Zarqa Ma'in, Wadi 162, 174, 211–12
Zarqa River 162, 192, 194
Zeboyim 43, 128
Zechariah (biblical prophet) 56
Ze'elim 8, 35–6, 56, 104, 162
Zenon papyri 73–4, 236
Zionism 148, 159, 171, 173–7, 183–5, 209
Zoar (Zoara) 2, 8, 25, 36, 42–4, 48, 86, 94, 97, 100, 102, 113–14, 116, 119, 128, 156
Zughar (medieval Zoar) 44, 104, 114–18, 123
Zwickel, Wolfgang 9, 29